Withdrawn
University of Waterloo

STOCHASTIC OPTIMIZATION AND ECONOMIC MODELS

THEORY AND DECISION LIBRARY

General Editors: W. Leinfellner and G. Eberlein

Series A: Philosophy and Methodology of the Social Sciences
Editors: W. Leinfellner (University of Nebraska)
G. Eberlein (University of Technology, Munich)

Series B: Mathematical and Statistical Methods
Editor: H. Skala (University of Paderborn)

Series C: Game Theory and Decision Making
Editor: S. Tijs (University of Nijmegen)

Series D: Organization and Systems Theory
Editor: W. Janko (University of Karlsruhe)

SERIES B: MATHEMATICAL AND STATISTICAL METHODS
Editor: H. Skala (Paderborn)

Editorial Board

J. Aczel (Waterloo), G. Bamberg (Augsburg), W. Eichhorn (Karlsruhe),
P. Fishburn (New Jersey), D. Fraser (Toronto), B. Fuchssteiner (Paderborn),
W. Janko (Karlsruhe), P. de Jong (Vancouver), M. Machina (San Diego),
A. Rapoport (Toronto), M. Richter (Aachen), D. Sprott (Waterloo),
P. Suppes (Stanford), H. Theil (Florida), E. Trillas (Madrid), L. Zadeh (Berkeley).

Scope

The series focuses on the application of methods and ideas of logic, mathematics and statistics to the social sciences. In particular, formal treatment of social phenomena, the analysis of decision making, information theory and problems of inference will be central themes of this part of the library. Besides theoretical results, empirical investigations and the testing of theoretical models of real world problems will be subjects of interest. In addition to emphasizing interdisciplinary communication, the series will seek to support the rapid dissemination of recent results.

JATI K. SENGUPTA

*Department of Economics, University of California,
Santa Barbara, U.S.A.*

STOCHASTIC OPTIMIZATION AND ECONOMIC MODELS

D. REIDEL PUBLISHING COMPANY

A MEMBER OF THE KLUWER ACADEMIC PUBLISHERS GROUP

DORDRECHT / BOSTON / LANCASTER / TOKYO

Library of Congress Cataloging in Publication Data

Sengupta, Jatikumar.
  Stochastic optimization and economic models.

(Theory and decision library. Series B, Mathematical and statistical methods)
Includes index.
1. Mathematical optimization. 2. Stochastic processes. 3. Decision-making. 4. Uncertainty. 5. Econometric models. 6. Portfolio management. I. Title. II. Series.
HB143.7.S46   1986   330′.01′5192   86–13880
ISBN 90–277–2301–X

Published by D. Reidel Publishing Company,
P.O. Box 17, 3300 AA Dordrecht, Holland.

Sold and distributed in the U.S.A. and Canada
by Kluwer Academic Publishers,
101 Philip Drive, Assinippi Park, Norwell, MA 02061, U.S.A.

In all other countries, sold and distributed
by Kluwer Academic Publishers Group,
P.O. Box 322, 3300 AH Dordrecht, Holland.

All Rights Reserved
© 1986 by D. Reidel Publishing Company, Dordrecht, Holland
No part of the material protected by this copyright notice may be reproduced or
utilized in any form or by any means, electronic or mechanical
including photocopying, recording or by any information storage and
retrieval system, without written permission from the copyright owner

Printed in The Netherlands

To my mother

       with devotion and love

Contents

Preface ix

I. Econometric Models and Optimal Economic Policy 1

    Problems of Specification 4
    Robustness Issues 6

II. Stochastic Processes in Economic Models 11

    Introduction 11
    Commonly Used Stochastic Processes 12
    Illustrative Applications 25

III. Recent Economic Models in Applied Optimal Control 53

    Introduction 53
    Conjectural Equilibria in Cournot-Nash Games 53
    Stabilization Policy in Cartelized Markets 83
    Adaptive Controls in Economic Models 97
    Renewable Resources Model under Uncertainty 110

IV. Efficient Diversification in Optimal Portfolio Theory 123

    Introduction 123
    Estimation Risk in Portfolio Theory 130
    Minimax Portfolio 143
    Decision Rules for Portfolio Revision 153
    Dynamic Rules of Revision 169
    Concluding Remarks 187

V. Portfolio Efficiency under Singularity and Orthogonality 190

    Introduction 190
    Mean Variance Model under Singularity 193
    A Distance Criterion Approach 218
    Dynamic Aspects 226
    Concluding Remarks 231

CONTENTS

VI. Diversification and Robustness in Portfolio Investment: An Empirical Analysis — 233

    Introduction — 233
    Diversification in Small Size Portfolios — 237
    Portfolio Performance of Mutual Funds — 267
    Evaluation of Mutual Fund Portfolios — 285
    Estimation of Portfolio Efficiency Frontier — 304
    Concluding Remarks — 314

VII. Efficiency Measurement in Nonmarket Systems Through Data Envelopment Analysis — 321

    Introduction — 321
    Nonparametric Methods of Efficiency Measurement — 325
    Generalizations of the DEA Model — 332
    An Empirical Application — 360

Author Index — 368

Subject Index — 371

Preface

This book presents the main applied aspects of **stochastic optimization in economic models**. Stochastic processes and control theory are used under optimization to illustrate the various economic implications of optimal decision rules.

Unlike econometrics which deals with estimation, this book emphasizes the decision-theoretic basis of uncertainty specified by the stochastic point of view. Methods of applied stochastic control using stochastic processes have now reached an exciting phase, where several disciplines like systems engineering, operations research and natural resources interact along with the conventional fields such as mathematical economics, finance and control systems. Our objective is to present a critical overview of this broad terrain from a multidisciplinary viewpoint. In this attempt we have at times stressed viewpoints other than the purely economic one. We believe that the economist would find it most profitable to learn from the other disciplines where stochastic optimization has been successfully applied. It is in this spirit that we have discussed in some detail the following major areas:

    A. Portfolio models in finance,

    B. Differential games under uncertainty,

    C. Self-tuning regulators,

    D. Models of renewable resources under uncertainty, and

E. Nonparametric methods of efficiency measurement. Stochastic processes are now increasingly used in economic models to understand the various adaptive behavior implicit in the formulation of expectation and its application in decision rules which are optimum in some sense. Our selected models are designed to evaluate such stochastic processes as Wiener process, jump process and diffusion process in their implications for (a) operational usefulness in policy making, (b) computational flexibility and (c) adaptivity to new information and data.

I have been involved in applied research on stochastic economics for over a decade, since my teacher Gerhard Tintner introduced me to this exciting field. He served as my mentor, philosopher and guide. This book bears a small testimony to my research indebtedness to him in this field which was so close to his heart.

Finally, I am most deeply in appreciation of the support of my wife, who never failed to believe in my capability, even when I was not so sure. It is indeed a real pleasure to record my appreciation of her along with my two children for their enduring support.

Jati K. Sengupta
Professor of Economics
   and Operations Research

University of California
Santa Barbara, California
April 1986

CHAPTER I

Econometric Models and Optimal Economic Policy

Specifying optimal economic policy using an econometric model has passed through three phases in its evolution. The first phase started with a two-stage approach in Tinbergen's tradition: first, one estimates an econometric model by the standard techniques of statistical estimation, then one uses optimal control theory in the estimated model to determine an optimal policy. This approach failed to recognize two types of interdependence. One is that the changes in policy or optimal control would induce changes in the structural response of the model. Thus the post-control model equations would differ from the pre-control ones, thus generating some inconsistency. The second is that future errors or uncertainty conditional on the adoption of a given set of policies would be very different from the past by the same reasoning. The second phase of econometric model building for specifying an optimal economic policy explicitly allows for the interdependence of the two conditional aspects: the problem of optimal estimation given the control or policy variables and the problem of optimal regulation or control, given the estimated model using all past information upto the current date. The research work by modern control theorists [1,2] on self-tuning and other adaptive controllers has been explicitly introduced in economic models. These attempts have served to clarify our understanding of some of the basic aspects of a stochastically optimal economic policy, whenever it exists. Some of these aspects can be qualitatively described in brief as follows:

1. The type of information available as of now and in future affect the specification of an optimal stochastic control very significantly,
2. Adaptivity of a controller as in the methods of self-tuning control may take several forms, depending on the elements of caution and learning incorporated, and
3. The adaptive decision rules which are optimal in some sense frequently involve significant nonlinearities as soon as the LQG (Linear quadratic and Gaussian) framework is generalized, thus leaving open the question whether the computational complexity required for fine-tuned optimal control is worth paying for [3].

The third phase of econometric policy modeling explicitly allows for a multiplicity of players each with his own set of control or policy variables. This yields the framework of a differential game, where the state vector is jointly influenced by the conjectures of any one player about others' potential strategies. In a non-cooperative framework two additional sources of inconsistency may arise here. One is that the players' subjective expectations and their successive revisions conditional on each outcome of the game as of time t may not converge, unless a process of implicit coordination or learning can be built into it. The second reason is the difficulty in an econometric or statistical sense to identify and test for a mixed-strategy equilibrium solution, when pure strategy solutions fail to exist. Suitable econometric tools are not yet available to test for mixed strategy vectors satisfying for example a Cournot-Nash equilibrium or other bargaining solutions.

It is thus clear that if the environment is fluctuating in a stochastic sense and the players have to choose their mixed strategies interdependently, the problem of specifying an optimal policy which is mutually consistent is not very simple anymore. Such situations do frequently arise in models of renewable resources like the fishing population in sea coasts which is harvested by different groups, each following procedures of random search. From a practical standpoint one could settle for some approximate and simple policies which may not be very fine tuned or optimal but possess some properties of robustness. It seems that in the framework of stochastic differential games, the characterization of simple yet robust policies may prove to be most valuable for the policy makers. As Johansen [4] has put the matter very succinctly:

"Consider a government which is about to choose between possible policies $a^{(1)}, a^{(2)}, \ldots, a^{(n)}$. It would be rather artificial to assume that the government could realistically say to itself: 'we have to choose between the various possible policies, but regardless of which decision we eventually settle for, our decision has already been anticipated by the rest of the economy'. Now the theory of rational expectations does not mean that anticipations are necessarily correct, so this way of putting the matter is somewhat too strict. However, the theory assumes the deviations between anticipations and realizations to be purely random. This modification does not help very much to make the viewpoint relevant for a government who is pursuing a discretionary policy. A government in such a position must, of course to the extent that anticipations influence behavior, assume something about the anticipations which the other decision-makers hold about the government's policy, but the theory of rational expectations does not give an adequate or complete formulation for such a situation (p. 19)".

1. <u>Problems of Specification</u>

Stochastic optimization in economic models may be explicit or implicit. It is explicit when separate variables or parameters are introduced specifically for the purpose e.g. risk aversion, cost of adjustment due to uncertainty, or a degree of pessimism or caution in the choice of policy in a stochastic world. Typical examples are provided by portfolio models of efficient investment, conjectural equilibria in Cournot-Nash market models and models of renewable and exhaustible resources under various conditions of uncertainty.

Implicit stochastic optimization involves either latent variables or, nonparametric decision rules. In multivariate statistics the principal components of a set of random variables provide an example of latent variables. An example of nonparametric decision rules is provided by the efficiency ranking and measurement by a suitable convex closure of a set of random points, which has been recently applied by a number of researchers [5-7]. The questions we ask in case of implicit stochastic optimization are very different from those of explicit stochastic optimization. For instance one major question asks: is the given data set consistent with a specific efficiency hypothesis? If not we have to consider then some transformations of the original data set, before efficiency in the sense of a production frontier say may be evaluated. It is clear that implicit stochastic optimization is more suitable in nonmarket systems like schools and public sector enterprises, where most of the inputs and outputs are observable in a limited way without any knowledge of their market prices. Charnes and Cooper [5] have used the term 'data envelopment analysis' (DEA) to include one class of implicit stochastic optimization models, where the shadow prices of observed inputs and realized outputs are derived by suitable models of linear and nonlinear programming. For applied economic models in the future, the implicit stochastic optimization yielding nonparametric measures or relationships are likely to be most valuable for two reasons. One is that the heterogeneity in the observed data structure in economics would be much more carefully assessed by the consistency tests employed in the DEA model. The second is that the issues of robustness would be directly admitted into the econometric testing procedures to an extent much greater than that which is currently practiced.

The models we have selected in this book to evaluate the role of stochastic optimization in applied economic models emphasize their possible usefulness and also limitations. Some of the useful applied aspects include the following:

(a) Comparisons with suitable deterministic models are made to evaluate the impact of uncertainty,
(b) Myopic and steady-state solutions are analyzed with their implications for stability, and
(c) The impact of the risk averse behavior is specifically evaluated in competitive and imperfectly competitive market structures.

2. Robustness Issues

Problems of robustness can be appreciated better in terms of the formal structure of a policy model, which is most often presented by a set of dynamic equations

$$f_t(x, z, \theta, \hat{\alpha}, \varepsilon) = 0 \tag{1}$$

where x is a vector of endogenous variables, z a vector of

control or decision variables, θ a set of parameters, $\hat{\alpha}$ a set of anticipated future values of any of the varialbes included in x and z and ε is a vector of unobservable random variables or disturbances which may enter additively or multiplicatively but in an explicit form. In order to define an optimal economic policy we need a preference function. A common practice is to maximize the expected value of an utility functional $U(x,z,\theta,\hat{\alpha},\varepsilon)$ yielding an objective function

$$\text{Max}_z J_T = E\{U(x,z,\theta,\hat{\alpha},\varepsilon)\} \qquad (2)$$

where T indicates a planning horizon. Assuming that an optimal solution in the form of a trajectory of the control vector exists, the impact of uncertainty on optimal policy may be evaluated in the following aspects:

(a) The vector $\hat{\alpha}$ of anticipations in the future needs to be forecast on the basis of all past information including the latest available. In the theory of rational expectations, $\hat{\alpha}$ is taken as the conditional forecast i.e. $\hat{\alpha} = E_t(x|I_1, I_2, \ldots, I_t)$ of the state variable given the information available, over the distribution of the future random variables $\varepsilon_t, \varepsilon_{t+1}, \varepsilon_{t+2}, \ldots$. Since these forecasts do not consider the higher moments above the mean and in case of nonnormal errors they may not have the best linear unbiasedness property, one may have to perform sensitivity analysis for the optimal policy trajectory.

(b) The parametric uncertainty enters through the problems of estimating the parameters θ and updating the optimal control trajectory as more data become available. As we have mentioned before optimal self-tuning controllers seek to combine optimal estimation with optimal control in a sequential fashion.

(c) Except for the class of control models known as the LQG models (i.e. linear quadratic and Gaussin), the optimal trajectory has to be solved from a nonlinear system of equations, which may exhibit fluctuations and oscillations. Robustness in the choice of economic policy then requires an adaptive policy which either smooths out these fluctuations or stays in a domain relatively free from them.

(d) In dynamic portfolio models when some of the parameters like θ i.e. the elements of the variance-covariance matrix of returns can not be directly estimated due to inadequate sample sizes, one may have to adopt caution by choosing a minimax policy i.e. a policy which minimizes the maximum loss; and finally

(e) in nonparametric models of efficiency measurement, the problem of robustness arises through the distribution of random observations around a stochastic production frontier.

These issues would be discussed in some detail in later chapters of the book. The main focus is on the applied aspects and hence likely to be most useful to the applied econometrician and the advisor to economic policy makers.

References

1. Astrom, K.J. and B. Wittenmark. On self-tuning regulators. Automatica 9 (1973), 185-200.

2. Sengupta, J.K. Information and Efficiency in Economic Decision. Dordrecht: Martinus Nijhoff Publishers, 1985.

3. Davis, M.H.A. and R.B. Vinter. Stochastic Modelling and Control. London: Chapman and Hall, 1985.

4. Johansen, L. Econometric models and economic planning and policy: some trends and problems. Memorandum of the Institute of Economics, University of Oslo, January 12, 1982.

5. Charnes, A. and W.W. Cooper. Management science relations for evaluations and management accountability. Journal of Enterprise Management 2 (1980), 143-162.

6. Sengupta, J.K. Data envelopment analysis for efficiency measurement in the stochastic case. Working paper, University of California, Santa Barbara, 1985.

7. Diewert, W.E. and C. Parkan. Linear programming tests of regularity conditions for production function, in Quantitative Studies on Production and Prices. Wurzburg, Austria: Physica-Verlag, 1983.

CHAPTER II

Stochastic Processes in Economic Models

I. Introduction

Stochastic processes have found increasing applications in modern economic models. In earlier times they were mainly used as additive errors or noise in a deterministic model without contributing very much to our basic understanding of the model structure, except perhaps hleping in providing a satisfactory basis of econometric estimation, e.g., the use of Cochrane-Orcutt estimation in autocorrelated errors. Recent economic applications have emphasized by contrast as it were three significant contributions of the theory of stochastic process in applied economic models:

(1) optimal control of dynamic stochastic systems: this leads to problems of stochastic control where the optimization objective has to be reformulated suitably,
(2) stochastic stability of dynamic models, where the parameters or their estimates may be stochastic: this leads to self-tuning regulators, where optimal estimation and optimal control may be suitably combined, and
(3) models of optimal search and learning in a random environment, where a suitable probabilitstic criterion, e.g., an entropy criterion has to be optimized: this leads to problems of improving the model specification which are also called problems of optimal design in control system engineering.

Our object here is to provide an introduction to some of the common types of stochastic processes and to illustrate their applications to economic and other fields.

CHAPTER II

2. <u>Commonly Used Stochastic Processes</u>

A stochastic process may be viewed in two ways. One is that it is simply a probability process; that is, any process in nature whose evolution can be analyzed successfully in terms of probability, i.e., the transition probability. A second view is that it is a family of random variables denoted by x_t or, $x(t)$ indexed by $t \, \varepsilon \, T$. The variation of t over the index set T introduces a dynamical element in the behavior of the random variable, which can be analyzed in two interrelated ways: either in terms of the probability functions dependent on t as t varies, or in terms of the expected value or variance functions of $x(t)$ as t changes. One useful way of division of stochastic processes is in terms of four classes, taking the case where we have one random variable at any particular t : (i) both x and t discrete, (ii) x discrete and t continuous, (iii) x continuous and t discrete and (iv) both x and t continuous. In each of the cases above one can define appropriate probability density functions. Stochastic processes can be classified in several ways depending on the criteria chosen. Three most common criteria are (a) the values taken by the random variable $x(t)$ being in the domain of positive integers or not, (b) the extent of dependence or independence of $x(t)$ on past history or memory and (c) the invariance or otherwise of the joint distribution function of the collection of random variables $x_1 = x(t_1), x_2 = x(t_2),\ldots,x_n = x(t_n)$ under an arbitrary translation of the time parameter.

If the process $x(t)$ is capable of assuming positive integral values only, then it is known as point processes, for in such a case, $x(t)$ can be identified with events or incidences represented as points on the time axis t . Although the parameter t is restricted to play the role of time, the theory of point processes has been generalized to include the multidimensional nature of the parameter t and has been widely applied in ecological problems. A very special class of such processes where the time intervals between successive events are independently and identically distributed are known as renewal processes, that have been most widely studied in statistical literature.

The second criterion introduces the classification of Markovian and non-Markovian processes. For any parametric value t and arbitrarily small $h > 0$, the probability structure of the process corresponding to the parametric value $(t+h)$ may depend only on the probability structure at t. In such a case $x(t)$ is called a Markov process; non-Markovian processes then consist of those processes which do not satisfy the above property of the Markov process known as the Markov property. An alternative way to define Markovian processes is through the joint distribution function of the random variables $X_1 = X(t_1)$, $X_2 = X(t_2),\ldots,X_n = X(t_n)$ for any finite set of $(t_1,t_2,\ldots,t_n)$ of $t \in T$, T being the index set which is denoted as

$$F_n(x_1,t_1;x_2,t_2;\ldots;x_n,t_n) = F_{X_1 X_2 \ldots X_n}(x_1,t_1;x_2,t_2;\ldots,x_n,t_n)$$

$$= P\{X_1 \leq x_1, X_2 \leq x_2,\ldots,X_n \leq x_n\} \tag{1.1}$$

where $x_i = x(t_i)$ and P denotes the joint probability. Thus the process $\{X(t), t \in T\}$ is Markovian, if for every n and for $t_1 < t_2 \ldots < t_n$ in the index set T we have

$$F\{x_n,t_n|x_{n-1},t_{n-1};x_{n-2},t_{n-2};\ldots\ldots;x_1,t_1\}$$

$$= F\{x_n,t_n|x_{n-1},t_{n-1}\} \ . \tag{1.2}$$

If the probability density function f of F exists, then (1.2) is equivalent to

$$f(x_n,t_n \mid x_{n-1},t_{n-1};x_{n-2},t_{n-2},\ldots\ldots,x_1,t_1)$$

$$= f(x_n,t_n \mid x_{n-1},t_{n-1})$$

On the basis of this relation an important equation known as the Chapman-Kolmogorov equation can be derived; several economic applications of this equation are discussed in Tintner and Sengupta [1], Mangel [2] and others [3]. Following the concept of memory in Markov property, one could define a memoryless process as one where the random variable $X(t)$ at a given t is independent of the random variables defined by $X(t)$ at all other t.

The invariance criterion classifies stochastic processes into stationary and nonstationary. A process $\{X(t), t \in T\}$ is defined to be strictly stationary if for each n and for every arbitrary τ we have for the distribution function $F_n(\cdot)$ defined in (1.1)

$$F_n(x_1,t_1;x_2,t_2;\ldots\ldots;x_n,t_n)$$
$$= F_n(x_1,t_1+\tau,x_2,t_2+\tau;\ldots\ldots;x_n,t_n+\tau) \qquad (1.3)$$

for $t_j + \tau \in T$, $j = 1,2,\ldots,n$ identically. Thus the family of joint probability distributions remains invariant under an arbitrary translation of time. Thus it follows that the distribution function $F_1(x,t)$ is not a function of t and

$$F_2(x_1,t_1;x_2,t_2) = F_2(x_1,x_2;t_2 - t_1) \quad .$$

It is clear from (11.3.2) that if the moments exist we would have

$$E\{X(t)\} = \text{constant, independent of time}$$
$$E\{X(t)\ X(t+\tau)\} = R(\tau) = R(-\tau) \quad .$$

A weaker concept than strict stationarity is provided by second-order sationarity. Thus a process $\{X(t), t \in T\}$ is wide-sense, or weakly or covariance, or second-order stationary if

$$E\{X(t)\} = \text{finite constant}$$

and

$$E\{X^2(t)\} < \infty, \quad E\{X(t_1) X(t_2)\} = R(t_2 - t_1) \quad .$$

Stationarity can also be defined by a process having a constant spectral density.

In economic applications Markov processes which are stationary in some sense have played very important parts. Reasons are several. First, the practicing econometricians had found very early that autoregressive systems with distributed lags fit very well several kinds of economic behavior, e.g., cobweb cycle, distributed lags, capital-stock adjustment. Second, use of the Chapman-Kolmogorov equation showed that differential equations representing dynamic behavior may be interpreted as dynamic random equations, of which the solutions or trajectories may be interpreted probabilistically. Since the solutions of random differential equations may or may not converge to their steady state equilibrum values, depending on certain conditions, one could obtain a generalized view of stochastic equilibrium or disequilibrium. Thus for instance, a deterministic differential equation may have a convergent solution as $t \to \infty$, while a stochastic analogue may not. Third, the Markovian property is very useful in modelling expectation formations and learning and even for non-Markovian situations the Markovian assumption is a good first approximation, especially when empirical data is very short.

One of the most useful applications of Markov processes is in terms of increment processes, defined over continuous time t. Consider the stochastic process $\{X(t)\}$ for $t \geq 0$. Denote the difference $X(t_2) - X(t_1)$ by $X(t_1,t_2)$ where $t_2 > t_1 > 0$, which is termed an increment of $X(t)$ on $[t_1,t_2]$. If for all $t_1 < t_2 < \ldots < t_n$, the successive increments $X(t_1,t_2), X(t_2,t_3), \ldots, X(t_{n-1},t_n)$ are mutually statistically independent, then the process $\{X(t), t \geq 0\}$ is called an <u>independent-increment</u> stochastic process. If for this process the probability distributions of its increments $X(t_1,t_2), X(t_2,t_3), \ldots, X(t_{n-1},t_n)$ depend only on the parameter differences $t_2 - t_1, t_3 - t_2, \ldots, t_n - t_{n-1}$ then the process $X(t)$ is said to have <u>stationary independent increments</u>. Three important examples of such processes which have found widemost applications are (a) the Brownian motion, named after its discoverer Robert Brown, a botanist; this is also called Wiener process or Wiener-Levy process that was widely applied by Einstein, (b) the Poisson process, which is most widely applied in queuing, telephone traffic and other fields of operations research and (c) Gaussian stochastic process, which is most frequently applied due to its analytic properties and the fact that the central limit theorem gives it a wider applicability.

More formally a Brownian process $\{X(t), t \geq 0\}$, which may now be denoted by $X(t,w)$, $t \varepsilon T$, $w \varepsilon W$ to indicate its dependence on the sample space W is a process satisfying four conditions: (i) $X(0,w) = 0$ by convention, i.e., the process starts at zero, (ii) stationary independent increments, (iii) the increment $X(t) - X(s)$, $t > s > 0$ has normal or Gaussian distribution with zero mean and variance $\sigma^2(t - s)$ and (iv) for each sample index $w \varepsilon W$ the process $X(t,w)$ is continuous in t for $t \geq 0$. The most significant contribution of Wiener on the development of the theory of Brownian motion is to show that the sample functions $X(t,w)$ viewed as functions of $w \varepsilon W$ are continuous but not differentiable.

The second example of an independent increment stochastic process is the Poisson process. The only difference here is that the process $X(t), t \geq 0$

has independent integer-valued increments. Arrival of customers in queue, sequence of production and inventories, spread of impulse noises are modelled by the Poisson process.

Lastly, a stochastic process $\{X(t), t \in T\}$ is called Gaussian, if for every finite set $t_1, t_2, \ldots, t_n$ the random variables $X(t_1), X(t_2), \ldots, X(t_n)$ have a joint nromal distribution with a mean vector $m'(t) = (EX(t_1), EX(t_2), \ldots, EX(t_n))$ and variance-covariance matrix $V = (v_{ij})$, where $v_{ij} = E\{(X(t_i) - m(t_i))(X(t_j) - m(t_j))\}$. In the stationary case the mean vector $m(t)$ is a constant (i.e., time-independent) and the covariance matrix is a function of the time instants only through their differences.

A concept associated with Gaussian processes that is most useful in interpreting random differential equation is the white noise property. Loosely speaking, a white noise process is defined by the derivative of a Brownian or Wiener process. But a Wiener process is not differentiable in the mean square sense, since by its definition given before we have

$$E\left(\frac{X(t) - X(s)}{t - s}\right)^2 = \frac{\sigma^2}{t - s} \to \infty \text{ as } \tau = (t - s) \to 0$$

hence one has to proceed differently. Let $W(t)$, $t \geq 0$ denote the Wiener process, which is by definition Gaussian with zero mean and covariance denoted by $\mu(t,s) = 2(t - s)\sigma^2$, where $\sigma^2 = c$ is a finite constant and $t > s$. The formal derivative of $W(t)$, denoted by $\dot{W}(t)$ is clearly Gaussian with mean zero and covariance given by

$$\begin{aligned}\mu_{\dot{W}(t)}(t,s) &= \partial^2 \mu(t,s)/\partial t \partial s \\ &= 2c\partial^2 \min(t,s)/\partial t \partial s \\ &= 2c\delta(t - s)\end{aligned} \quad (1.4)$$

$$\text{where } \delta(t - s) = \begin{cases} 0, & t < s \\ 1, & t > s \end{cases}.$$

CHAPTER II

The most important use of Gaussian white noise processes is in the theory of random differential equations, where Ito's theorem has to be used for taking account of the nondifferentiability property of the white noise process. Taking the linear scalar case we have the dynamic system

$$\dot{x}(t) = a(t) x(t) + b(t) w(t); \quad x(t_0) = x_0 \qquad (1.5a)$$

where $a(t)$, $b(t)$ are nonrandom time-dependent coefficients and $w(t)$ is Gaussian white noise independent of x_0. One could write (1.5a) alternatively as

$$dx(t) = a(t) x(t) dt + b(t) dn(t), \quad x(t_0) = x_0 \qquad (1.5b)$$

where we have used $dn(t)/dt = W(t)$

or, on integration

$$x(t) - x(t_0) = \int_{t_0}^{t} a(s) x(s) ds + \int_{t_0}^{t} b(s) dn(s) \qquad (1.5c)$$

$$x(t_0) = x_0 .$$

Whereas the first integral on the right-hand side of (1.5c) is well defined as Riemann integral, the second one may not, since the random variable y_n defined by the partial sum

$$y_n = \sum_{i=1}^{k} x(t_i)(n(t_i) - n(t_{i-1})) \qquad (1.5d)$$

does not converge in the mean square sense to a unique limit. Ito's theorem provided a method of selecting the subdivisions t_i, $t_i - t_{i-1}$ so that the limit of y_n in (11.5.4) exists as $\Delta_m \to 0$ where $\Delta_m = \max_i (t_{i+1} - t_i)$.

For specific applications of Ito's rules of stochastic calculus to any functional $z(t,x(t))$ defined on $x(t)$ satisfying (1.5b) we proceed as follows: we define the differential of $Z(t) = z(t,x(t))$ as

$$dZ(t) = [z_t + z_x a(t) + \tfrac{1}{2} z_{xx}(b(t))^2] \, dt + z_x b(t) \, dn(t) \qquad (1.5e)$$

where $z_t = \partial z(t,x(t))/\partial t$, $z_x = \partial z(t,x(t))/\partial x$ and $z_{xx} = \partial^2(t,x(t))/\partial x^2$. This differential exists under certain regularity conditions. In general if $Z(t) = z(t,x_1(t),x_2(t),\ldots,x_m(t))$ is a continuous function with continuous partial derivatives z_t, z_{x_i} and $z_{x_i x_j}$ and we have in place of (1.5b) the stochastic differentials on $[t_0,T]$

$$dx_i(t) = a_i(t) \, dt + b_i(t) \, dn(t), \qquad i=1,2,\ldots,m$$

then $Z(t)$ also possesses a stochastic differential in the same interval given by

$$dZ(t) = z_t \, dt + \sum_{i=1}^{m} z_{x_i} \, dx_i + \tfrac{1}{2} \sum_{\substack{i=1 \\ j=1}}^{m} z_{x_i x_j} \, dx_i \, dx_j \qquad (1.5f)$$

where $dx_i \, dx_j = b_i(t) \, b_j(t) \, dt$, $i,j \leq m$

Some examples of applying the Ito differential rules are as follows: Let $Z(t) = x_1(t) \, x_2(t)$ and

$$dx_1(t) = a_1(t) \, dt + b_1(t) \, dn(t)$$
$$dx_2(t) = a_2(t) \, dt + b_2(t) \, dn(t)$$

where $dn(t) = W(t) dt$ is the Gaussian white noise. On applying (1.5f) one gets

$$d(x_1(t) x_2(t)) = x_1(t) dx_2(t) + x_2(t) dx_1(t) + b_1(t) b_2(t) dt$$

$$= [x_1(t) a_2(t) + x_2(t) a_1(t) + b_1(t) b_2(t)] dt$$

$$+ [x_1(t) b_2(t) + x_2(t) b_1(t)] dn(t)$$

Take another scalar case where $Z(t) = z(x_0, e^{at})$ where a is a constant and $dx(t) = ax(t) dt$. Then $dZ(t) = z_t dt + z_x dx + \frac{1}{2} z_{xx} dx dx = ax_0 e^{at} dt$.

By applying these rules the solution of the linear scalar random differential equation (1.5b) taken in a more general form as

$$dx(t) = a(t) x(t) + \sum_{i=1}^{m} b_i(t) x(t) dn_i(t) \qquad (1.6a)$$

$$x(t_0) = c, \quad dn(t) = (dn_i(t)) = \text{m-dimensional white noise}$$

can be explicitly written as

$$x(t) = c \exp [\int_{t_0}^{t} (a(s) - \sum_{i=1}^{m} b_i(s)^2/2) ds + \sum_{i=1}^{m} \int_{t_0}^{t} b_i(s) dn_i(s)]$$

of $a(t), b_i(t)$ are constants, then we have

$$x(t) = c \exp [(a - \sum_{i=1}^{m} b_i^2/2)(t - t_0) + \sum_{i=1}^{m} b_i(n_i(t) - n_i(t_0))] \qquad (1.6b)$$

This result (1.6b) is very useful for application to economic models. For an example consider the proportional feedback rule of government policy analyzed by Phillips [4] under a Keynesian model

$$dy(t) = ay(t) + dg(t) = ay(t) + by(t) dn(t)$$

where $dy(t)$ is deviations of real income from the target level, the government policy intervention $g(t)$ is of the form $dg(t) = by(t)\,dn(t)$, with $dn(t)$ being a white noise Gaussian process and a, b are real constants. The income solution $y(t)$ is now given by

$$y(t) = c \exp[(a - b^2/2)(t - t_0) + b(n(t) - n(t_0))] .$$

Assuming that the law of large numbers holds for the $n(t)$ process, we would have $\lim y(t) \to 0$, as $t \to \infty$ only if $a < b^2/2$; otherwise the controlled system would be unstable. In his original work [4] Phillips argued in terms of only a deterministic differential equation system and showed that in many cases the government policy itself may be destabilizing. Here we find additional stochastic reasons why this may be so.

It is clear that in discrete-time cases the commonly used stochastic processes present no additional problems over and above the presence of randomness and the fact that there may be nonstationarity over time. But in continuous-time cases there arise difficulties due to the fact that a stochastic process may not have a derivative in the strict sense of the term. Two examples will clarify this point. One is a Brownian motion process $W(t)$ that is completely continuous in time and the other is a jump process which has completely discontinuous time-paths.

A Brownian motion process $W(t)$ has the following properties

(i) sample paths of $W(t)$ are continuous,

(ii) $W(0) = 0$,

(iii) the increment $W(t+s) - W(t)$ is normally and independently distributed with mean zero and variance $t\sigma^2$, where σ^2 is a fixed constant.

CHAPTER II

Letting $dW = W(t+dt) - W(t)$, the third property implies that $E\{dW\} = 0$ and $E\{(dW)^2\} = \sigma^2 dt$. The fact that the variance of dW is of order dt, creates some mathematical difficulties, since for example we have

$$D\{(dW/dt)^2\} \sim \sigma^2/dt \to \infty \text{ as } dt \to 0$$

This implies that $W(t)$ has not had a derivative in a strict mathematical sense, i.e., a Brownian motion is everywhere continuous but nowhere differentiable.

The derivative $\varepsilon(t) = dW/dt$ is called Gaussian white noise as we have seen before. On using the correlation function

$$E\{\varepsilon(t)\ \varepsilon(s)\} = \frac{\partial^2 E\{W(t)\ W(s)\}}{\partial t \partial s}$$

$$= \frac{\partial^2}{\partial t \partial s} \min(t,s)\ \sigma^2 = \delta(t-s)\ \sigma^2, \qquad (2.1)$$

where $\delta(t-s)$ is the delta function:

$$\delta(t-s) = \lim_{n \to \infty} (n/\sqrt{2\pi})\ \exp[-n^2(t-s)^2/2]$$

We may easily derive the transition probability density as

$$\text{Prob }\{W(t) \in (x, x+dx)\ W(s) = y\}$$

$$= \{2\pi\sigma^2(t-s)\}^{-\frac{1}{2}} \exp\{-\frac{(x-y)^2}{2\sigma^2(t-s)}\}\ dx, \qquad (2.2)$$

Two most common applications of Brownian motion process are the following: one is to use the transiton probability (2.2) and how it changes under different forms of feedback control, e.g., derivative, proportional or integral [1,4]. The second is to use it as a stochastic diffusion process which underlies the theory of stochastic differential equations. Since the explicit solutions of stochastic differential equations are not always explicitly computable in a closed form, various approximation methods are applied. One most frequently used method of approximation is the method of moments; it asks if it is possible from the information on the stochastic process X(t) which satisfies a stochastic differential equation to compute its mean and variance functions.

Now consider a jump process $\pi(t)$ which satisfies the following assumptions:

(i) $\pi(0) = 0$,

(ii) Given that $\pi(t) = x$, the waiting time until the next jump has a one-parameter exponential distribution with expectation $1/\alpha(x)$,

(iii) Given that a jump occurs, then the probabilities of the increment $\Delta\pi$ and decrement $-\Delta\pi$ are:

$$\text{Prob }\{\Delta\pi = a\} = \lambda(x), \text{ Prob }\{\Delta\pi = -b\} = \mu(x)$$

where a,b are positive numbers

(iv) and the jumps are independent of the waiting times.

These assumptions characterize a Markovian process which is widely used in queuing and waiting line models under the name of stochastic birth and death process. For example assume $a = b = 1$ and that $\pi(t)$ can take only integer values $x = 0, 1, 2, \ldots, n, n+1, \ldots$. Let us denote

$$P_n(t) = \text{Prob }\{\pi(t) = n\}, \text{ with } \pi(0) = 0 \qquad (2.3)$$

CHAPTER II

then the four assumptions above lead to the difference-differential equation

$$P_n(t+dt) = P_n(t)\{1 - (\lambda_n+\mu_n) dt\}$$
$$+ (\lambda_{n-1}dt) P_{n-1}(t) + (\mu_{n+1}dt) P_{n+1}(t)$$
$$+ 0(dt) \quad (2.4)$$

where $0(dt)$ denotes those remaining terms becoming negligible as $dt \to 0$. Letting $dt \to 0$ we obtain a set of mixed linear differntial equations

$$\frac{dP_n(t)}{dt} = -(\lambda_n+\mu_n) P_n(t) + \lambda_{n-1}P_{n-1}(t) + \mu_{n+1}P_{n+1}(t) \quad (2.5)$$

where $\lambda(n) = \lambda_n$, $\mu(n) = \mu_n$ and it is assumed

$$P_0(0) = 1, P_n(0) = 0 \text{ for } n \geq 1.$$

If the so-called birth (λ_n) and death (μ_n) rates are positive constants, i.e., $\lambda_n = \lambda$, $\mu_n = \mu$, then a steady state solution of (2.5) can be easily obtained as

$$P_n = \theta^n(1-\theta), \; \theta = \lambda/\mu \quad (2.6)$$

on using which the mean and variance functions of $\pi(t)$ can be determined. Thus if $P_n(t)$ is the complete solution of (2.5), the mean $M(t)$ and variance functions $V(t)$ are

$$M(t) = \sum_{n=0}^{\infty} nP_n(t), \; V(t) = \sum_{n=0}^{\infty} [n-M(t)] P_n(t)$$

Two types of applications are commonly made for these types of Markovian models. One is in queuing, where there are several channels of servicing customers who arrive randomly [5]. The main interest here is to study the various characteristics of the queuing process, e.g., mean waiting time, average length of the queue, the rate of increase of $M(t)$ or $V(t)$ for different subintervals of time. A second is the control theoretic application. Let $\theta = \theta(\lambda,\mu)$ be subject to some decision or control variables in say the number of channels of service, the rate of promotional spendings etc. In this case the transition probabilities $P_n(t)$ and therefore the mean-variance functions $M(t)$, $V(t)$ become functions of the control variables $u(t)$. One could then optimize a suitable performance measure of the Markovian system by choosing the control variables appropriately.

3. Illustrative Applications

We may now consider some examples of application of selected stochastic processes to economics and other operational fields. These examples illustrate the following problems in stochastic processes and control:

A. Optimal control of a queuing process,
B. Stochastic control as Kalman filter problem,
C. Optimal extraction of a natural resource under demand uncertainty,
D. Problems of optimal search, and
E. Stability of stochastic systems satisfying dynamic random equations.

Example 1 (Optimal control of a queuing process)

A queue is a waiting line of customers requiring service and it develops whenever the current arrival of customers exceeds the current capacity of the

CHAPTER II

service facility. Since the arrival and service processes are both random being subject to a stochastic process, a queue is most likely to develop at some time in the queuing system.

We use a standard queuing system with exponential inter arrival and service time distributions where linear birth and death processes discussed in (2.5) and (2.6) apply. There will be S servers and it is assumed that the maximum queue size will not exceed $N \geq S$, $S \geq 1$. The service time rate $u(t)$ is chosen by minimizing the cost of expected queue length in the system plus the cost of service over a specified time interval. Such models have been analyzed extensively in operations research [1,6]. Denote by $x_n(t)$ the probability of n customers in the system at time t, where customers arrive at Poisson times with mean arrival rate $\lambda(t)$, then the transition probabilities $\dot{x}_n(t) = dx_n(t)/dt$ are given as:

$$\dot{x}_0(t) = -\lambda(t) x_0(t) + u(t) x_1(t), \quad n = 0$$

$$\dot{x}_n(t) = \lambda(t) x_{n-1}(t) - [\lambda(t) + nu(t)] x_n(t)$$
$$+ (n+1) u(t) x_{n+1}(t) \quad 1 \leq n \leq S-1 \quad (3.1)$$

$$\dot{x}_n(t) = \lambda(t) x_{n-1}(t) - [\lambda(t) + Su(t)] x_n(t)$$
$$+ Su(t) x_{n+1}(t), \quad S \leq n \leq N-1$$

$$\dot{x}_n(t) = \lambda(t) x_{n-1}(t) - Su(t) x_N(t), \quad n = N$$

with the initial condition $\sum_{n=0}^{N} x_n(0) = 1$

The objective functions is to minimize J:

$$\min_{u(t)} J = \int_0^T \{ \sum_{n=0}^{N} nx_n(t) + c(u(t)) \} \, dt$$
$$+ \frac{1}{2} b \{ \sum_{n=0}^{N} nx_n(T) - k \}^2 \quad (3.2)$$

where k is the desired final time expected queue length, b is the relative cost of deviating from the desired level k and $c(u(t))$ is the service rate cost assumed to be strictly convex and increasing, i.e., $c(0) = 0$, $c'(u(t)) > 0$ and $c''(u(t)) > 0$ for $u(t) > 0$. One may also adjoin a feasibility constraint on $u(t)$ with a fixed upper bound $\bar{u}$:

$$0 \leq u(t) \leq \bar{u}, \quad t \in [0,T] \tag{3.3}$$

One could now apply Pontryagin's minimum principle to derive the optimal service process trajectory. For example let us assume $N = S = 1$, $c(u) = \frac{1}{2} au^2$, a being a positive constant. Then we have the optimal control problem

$$\min J = \tfrac{1}{2} b[x_1(T) - k]^2 + x_2(T)$$

subject to $\dot{x}_0 = -\lambda x_0 + u x_1$, $x_0(0) = x_{00}$

$$\dot{x}_1 = \lambda x_0 - u x_1, \quad x_1(0) = x_{10} \text{ (fixed)}$$
$$\dot{x}_2 = x_1 + \tfrac{1}{2} au^2, \quad x_2(0) = 0$$
$$0 \leq u \leq \bar{u}$$

where $\lambda(t) = \lambda$ (constant) and

$$x_{n+2}(t) = \int_0^t \{ \sum_{n=0}^{N} n x_n(s) + c(u(s)) \} \, ds$$

so that

$$\dot{x}_2 = x_1 + \tfrac{1}{2} au^2, \quad x_2(0) = 0$$

CHAPTER II

On using $x_0 + x_1 = 1$ and $\bar{u} = 1$ the optimal control becomes

$$\min J = \tfrac{1}{2} b[x_1(T) - k]^2 + x_2(T)$$

subject to

$$\dot{x}_1 = \lambda(1-x_1) - ux_1, \quad x_1(0) = x_{10}$$
$$\dot{x}_2 = x_1 + \tfrac{1}{2} au^2, \quad x_2(0) = 0$$
$$0 \leq u \leq 1$$

By applying Pontryagin's minimum principle with adjoint variables z_1, z_2 to the Hamiltonian

$$H = z_1[\lambda(1-x_1) - ux_1] + z_2(x_1 + \tfrac{1}{2} au^2)$$

We see that w_2 is a constant equal to one say, since $\dot{w}_2 = -\partial H/\partial x_2 = 0$. The optimal trajectory must therefore satisfy:

$$\dot{z}_1 = z_1(\lambda+u) - 1, \quad z_1(T) = b[x_1(T) - k]$$
$$\dot{x}_1 = \lambda(1-x_1) - ux_1, \quad x_1(0) = x_{10} \qquad (3.3)$$
$$\dot{x}_2 = x_1 + \tfrac{1}{2} au^2, \quad x_2(0) = 0$$

$$\partial H/\partial u = -z_1 x_1 + au = 0 \text{ for } 0 \leq u \leq 1$$

hence

$$u = \begin{cases} 0, & \text{if } z_1 x_1/a < 0 \\ z_1 x_1/a, & \text{if } 0 \leq z_1 x_1/a \leq 1 \\ 1, & \text{if } z_1 x_1/a > 1 \end{cases}$$

The complete characterization of the trajectory can be very easily worked out.

Two points may be noted in this example. One is that the arrival rate $\lambda(t)$ itself may also be influenced by suitable control variables, e.g., special discount rates for certain arrival times. The second is that this type of model is typically applicable to models [5] of advertising which affect different brands of a product like cigarettes and to models of optimal inventory where things unsold arrive at rates $\lambda(t)$. Problems of estimation of the parameters λ in the transition probability equations also raise some interesting issues in dynamic econometrics.

Example 2 (Kalman filter in stochastic control)

Consider a single output (y_t) and single control (u_t) model with additive errors ε_t, where $\{\varepsilon_t\}$ is assumed to be a zero-mean wide sense stationary process with an uncorrelated sequence

$$A(z) y_t = B(z) u_t + \varepsilon_t \tag{4.1}$$

where the polynomials $A(z)$, $B(z)$ are lag operators defined as

$$A(z) = 1 + a_1 z^{-1} + \ldots + a_n z^{-n}$$

$$B(z) = b_1 z^{-1} + \ldots + b_n z^{-n}$$

Define

$$\beta^T = (b_1, \theta^T), \quad \theta^T = (b_2, b_3, \ldots, b_n, a_1, a_2, \ldots, a_n)$$

Then we obtain from (4.1)

CHAPTER II

$$y_t = \psi_t^T \theta + b_1 u_{t-1} + \varepsilon_t \qquad (4.2)$$

where $\psi_t^T = (u_{t-2}, \ldots, u_{t-n}, -y_{t-1} - \ldots - y_{t-n})$. By minimizing the one-period-ahead expected loss function

$$L = E\{y_{t+1}^2\}$$

we obtain the minimum variance (MV) control u_t^* as:

$$u_t^* = -\psi_{t+1}^T \theta / b_1 \qquad (4.3)$$

In order to implement this MV control u_t^* we have to know the parameter values (θ, b_1). If these are not known we have to replace them by their current estimate, e.g., by least squares (LS) principle, based on all past information. If the estimate is done by a recursive process over time, we obtain a two-stage procedure known as passive adaptive control. Various types of such procedures have been discussed in some detail in the recent control literature [5,7,8]. Let us consider some of the useful ones, e.g., (a) active adaptive control, (b) self-tuning regulator, and (c) an optimal design control.

In active adaptive control we consider the least squares (LS) estimate for the column vector $\beta = (b_1, \theta^T)^T$ in (4.2) recursively computed as follows:

$$\hat{\beta}_{t+1} = \hat{\beta}_t + K_{t+1}(y_{t+1} - x_{t+1}^T \hat{\beta}_t) \qquad (4.4)$$

$$K_{t+1} = (P_t x_{t+1})[1 + x_{t+1}^T P_t x_{t+1}]^{-1} \qquad (4.5)$$

$$P_{t+1} = [I - P_t \frac{x_{t+1} x_{t+1}^T}{1 + x_{t+1}^T P_t x_{t+1}}] P_t \qquad (4.6)$$

where $x_t^T = (y_{t-1}, \psi_t^T)$, the superscript T denotes the transpose and we can start the iterative process by using an estimate P_0 from initial data in (4.6) or by letting $P_0 = 1/h$ where h is small. As t gets large the choice of h is unimportant. Here (4.4) defines a Kalman filter which can be recursively computed as more and more data are available. Note that the Kalman filter (4.4) has three important features. The first is that it provides the new estimate $\hat{\beta}_{t+1}$ as a combination of the old estimate $\hat{\beta}_t$ and new observations on output, i.e.,

$$\hat{\beta}_{t+1} = [I - K_{t+1} x_{t+1}^T] \hat{\beta}_t + K_{t+1} y_{t+1}$$

Second, one can derive the recursive LS estimates (4.4) through (4.6) by minimizing a quadratic loss function

$$S_{t+1}(\beta) = S_t(\beta) + (y_{t+1} - x_{t+1}^T \beta)^2 \qquad (4.7)$$

with $\qquad S_t(\beta) = (\beta - \hat{\beta}_t)^T P_t^{-1} (\beta - \hat{\beta}_t) + \beta_t.$

It is clear that this method can handle time-varying parameters and if the gain matrix K_t converges to a steady state value, then this recursive LS method also converges. The third feature lies in its optimal method of discarding past data as new observations become available. It can also handle the situation when the errors of ε_t in (4.1) are autocorrelated. For in this case if the autocorrelation function is known, i.e.,

$$H(z) \varepsilon_t = \xi_t$$

where ξ_t is an uncorrelated white noise sequence and H(z) is a polynomial operator with an inverse which makes the autocorrelated errors ε_t uncorrelated.

CHAPTER II

The transformation of (4.1) then becomes

$$A(z) y_t^* = B(z) u_t^* + \xi_t \qquad (4.8)$$

with $\quad y_t^* = H^{-1}(z) y_t \quad$ and $\quad u_t^* = H^{-1}(z) u_t.$

We may then apply the Kalman filter updates to the transformed model (4.8).

Self-tuning regulators are those which combine the minimum variance controller and the recursive LS estimator. It is clear that methods of combining these two features are not very trivial. For one thing, any such method would depend for its performance on the uncertainties associated with the parameters, e.g., a Bayesian approach is one possibility. Secondly, if the errors ε_t in (4.1) are autocorrelated without their correlation structure known it is not always possible to guarantee the convergence of the two-stage procedure.

A more direct procedure is to apply the method of optimal design to obtain an active adaptive control as Goodwin and Payne [7] have done for the system (4.2) as follows. They specify a generalized cost function as

$$J = E_{y_{t+1}|\beta} \{y_{t+1}^2\} - \lambda(\det P_t / \det P_{t+1}) \qquad (4.9)$$

and determine an optimal u_t^* recursively by minimizing J, where $\det P_t$ is the determinant of the matrix P_t defined in (4.6) and λ is a positive weight preassigned. It is easy to derive the optimal control u_t^* as

$$u_t^* = - [b_{1t} \hat{\theta}_t^T \psi_{t+1} - \lambda w_t^T \psi_{t+1}][\hat{b}_{1t}^2 - \lambda s_t]^{-1}$$

where

$$P_t = \begin{pmatrix} s_t & w_t^T \\ w_t & Q_t \end{pmatrix}$$

is partitioned corresponding to the partition of β into b_1 and θ. The logic behind the cost function is simple: the first term is for providing a good control ($\lambda=0$), while the second term is aimed at making P_{t+1} small relative to P_t, i.e., error variance minimizing.

Example 3 (Optimal resource extraction under uncertainty)

We consider a problem of optimal management of an exhaustible (or non-renewable) resource, where we consider a simple case first where price p(t) per unit is exogenous nd deterministic but the stock of reserves R(t) at time t is assumed to follow a Brownian motion process

$$dR = -q\, dt + \sigma\, dW, \quad R(0) = R_0 \text{ (known)} \tag{5.1}$$

where W(t) in dW is a Brownian process, $q = q(t)$ is the rate of extraction, σ is a positive constant, and the initial reserve R_0 is assumed known exactly. The optimal resource extraction problem for a producer may then be set up as one of choosing q(t) so as to maximize the expected present value of future profits, i.e.,

$$\max_q E\{\int_0^T \pi(t)\, q(t)\, e^{-\delta t}\} \tag{5.2}$$

where

$$\pi(t) = p(t) - C(R(t))$$

$C(R)$ = cost of production when the stock level is $R = R(t)$.

With a fixed horizon [0,T], the optimal trajectory must satisfy the Euler-Lagrange necessary conditons of optimality

$$\frac{\partial F}{\partial r} = E\{\frac{d}{dt}(\partial F/\partial \dot{r})\}; \quad F = \pi(t)\, q e^{-\delta t}$$

where $R(t) = r$ and using (5.2) this can be written as

$$\frac{\partial F}{\partial r} = -\frac{1}{dt} E\{d(\frac{\partial F}{\partial q})\} \quad . \tag{5.3}$$

This yields

$$C'(r) qe^{-\partial t} = \frac{1}{dt} E\{-\delta(p(t) - C(R)) e^{-\delta t} dt + e^{-\delta t} dp - e^{-\delta t} dC(R)\}. \tag{5.4}$$

If the cost function $C(r)$ is strictly convex and twice differentiable, then one can write

$$dC(r) = C'(R) dR + \tfrac{1}{2} C''(R)(dR)^2 + o(dt)$$

and by (5.2)

$$E\{dR\} = -q \, dt, \quad E\{(dR)^2\} = \sigma^2 dt.$$

Thus the conditional expectation of dC is

$$E\{dC(r) \mid R(t) = r\} = -qC'(r) dt + \tfrac{1}{2} C''(r) \sigma^2 dt + o(dt). \tag{5.5}$$

On using this result in (5.4) we finally obtain

$$\frac{1}{dt} E\{dp\} = \delta E\{p(t) - C(R)\} + \tfrac{1}{2} \sigma^2 C''(r).$$

If $C''(r)$ is positive as is most likely to be in most cases, then it is clear that uncertainty in the form of $\sigma^2 > 0$ would tend to increase the rate of movement of the price of the resource.

Now consider a discrete-time case with a more general objective function

$$\max_{q_t} J = \sum_{t=0}^{\infty} [\delta^t EU(\tilde{\pi}(q_t))] \tag{6.1}$$

Subject to resource availability:

$$\sum_{t=0}^{\infty} q_t \leq R_0; \quad q_t, R_t \geq 0 \tag{6.2}$$

$$\delta = 1/(1+\rho), \quad \rho = \text{rate of discount (given)}$$

Here $U(\cdot)$ is the utility function with positive marginal utility $U' > 0$ and $U'' \gtreqless 0$ according as the resource owners are risk preferring, risk-neutral or risk-averse; $\tilde{\pi}(q_t)$ is random profits defined as

$$\tilde{\pi}(q_t) = q\tilde{p}(q) - C(q) \tag{6.3}$$

where

$$\tilde{p}(q_t) = \bar{p}(q_t) + \tilde{\varepsilon}_t \quad : \quad \text{random market price}$$

$$C(q_t) = \text{cost of extraction (nonrandom)}$$

with positive increasing marginal cost, i.e.,

$$C' > 0, \quad C'' > 0.$$

Here $\bar{p}(q_t) = \bar{p}$ is the expected price dependent on q_t and $\tilde{\varepsilon}_t$ is a random variable independently distributed with $\tilde{\varepsilon}_t \geq 0$ and $E(\tilde{\varepsilon}_t) = \varepsilon_0$. The assumption of non-negativity of $\tilde{\varepsilon}_t$ is introduced to prevent the possibility of negative market prices. The optimal extraction model above is due to Lewis [9] and similar models for commercial fisheries and other renewable resources have been formulated by Andersen [10] and others [11].

On using the Lagrange multiplier λ for the resource availability constraint (6.2) and applying the Kuhn-Tucker necessary conditons of optimality we obtain for the optimal extraction profile $\{q_t\}$:

CHAPTER II

$$\delta^t E\{'\tilde{\pi}_q\} - \lambda \leq 0 \text{ for } q_t \geq 0 \tag{6.4}$$

where $\tilde{\pi}_q = \partial\tilde{\pi}/\partial q$. These necessary conditions are also sufficient for (6.1) if it holds

$$E[U''\tilde{\pi}_q^2 + U''\tilde{\pi}_{qq}] < 0. \tag{6.5}$$

It is clear that if $\tilde{\pi}$ is strictly concave in q and U is either risk-averse or risk-neutral, i.e., $U'' \leq 0$, the sufficiency conditon (6.5) would hold. Assuming an interior soluton ($q_t > 0$) and the sufficiency condition (6.5) we obtain therefore the optimality rule

$$[\delta^t E(U'\tilde{\pi}_q) - \lambda] \, q_t = 0 \tag{6.6}$$

This says that along the optimal path increases in expected current profits are exactly offset by a decrease in future returns due to current resource use.

To compare the resource extraction behavior of risk averse and risk-neutral owners we may postulate a specific form of the utility function, e.g., a class exhibiting a constant rate of absolute risk-aversion: $-U''/U' = k$, k being a positive constant and then rewrite

$$\begin{aligned} EU(\tilde{\pi}) &= E\tilde{\pi} - \frac{k}{2}\sigma_\pi^2 \\ &= \bar{p}q - C(q) - \frac{k}{2}\sigma_\varepsilon^2 q^2 \\ &= (a - bq)q - C(q) - \frac{k}{2}\sigma_\varepsilon^2 q^2 \end{aligned} \tag{6.7}$$

where we have assumed a linear expected demand function $\bar{p} = a - bq$ and σ_ε^2 denotes the variance of $\tilde{\varepsilon}_t$ assumed to be a positive constant. Assuming interior solutions we obtain for the optimal extraction profile $q = \{q_t\}$:

$$q = (2b + k\sigma_\varepsilon^2)^{-1} \{a - C' - \lambda\delta^{-t}\}, \quad q > 0.$$

This yields $q(0) > q(k)$ for all positive k and $q(k_1) > q(0)$ for $k = -k_1$, $k_1 > 0$. It is clear that the risk preferring resource owners deplete more rapidly than the risk-neutral owners, who in turn deplete more rapidly than the risk averse owners.

One could interpret this model slightly differently by assuming that the resource owners are competitive, i.e., price takers when $\bar{p}$ is not dependent on q. Then on writing the Lagrangian function as

$$L = [\bar{p}q - C(q) - \frac{k}{2}\sigma_\varepsilon^2 q^2 + \tilde{\lambda}(R_0 - \sum_{t=0}^{\infty} q_t)] e^{-\delta t}$$

we obtain for the interior optimal solution

$$q = q_c = (k\sigma_\varepsilon^2)^{-1} [\bar{p} - C' - \hat{\lambda}] \qquad (6.8)$$

where $\hat{\lambda}$ is positive for the condition $R_0 = \Sigma q_t$ to hold. For the monopoly resource owner the condition analogous to (6.8) is

$$q = q_M = (k\sigma_\varepsilon^2)^{-1} [\bar{p}(1-\frac{1}{e}) - C' - \hat{\lambda}]$$

where e is the price elasticity of the expected demand curve. Thus for any $k \geq 0$ we get $q_M < q_c$ for all $e > 1$. But since a competitive resource owner is most likely to be risk-neutral, whereas a monopolist is likely to be risk-averse, the gap $(q_c - q_M)$ may still be higher.

CHAPTER II

Example 4 (Optimal search of resource deposits)

Exploration for an exhaustible resource like oil or mineral deposits involves random search and the cost of exploration depends therefore on the manner of search and the underlying probability structure of the natural deposit itself and that of the search procedure. Consider first a simple case of only one resource deposit in a given region A say. Let q(a) be the probability that the deposit has not been found after an area a of the region has been explored. Let (k/A) da be the probability of discovering the resource deposit if the area da is explored, where k is assumed to be a known constant. Then by the law of total probability we get

$$q(a + da) = q(a)[1 - (k/A) da].$$

This yields a differential equation in q:

$$\frac{dq}{da} = -\frac{k}{A} q \qquad (7.1)$$

with solution

$$q(a) = \exp\{-a(k/A)\} \qquad (7.2)$$

where $q(0) = 1$ is assumed. Suppose the area a is explored at a constant rate. Then set $ka/A = \beta t$ where β is a positive constant and t is time, i.e., time and area are linearly related. Then (7.2) becomes after setting $\hat{q}(t) = q(a)$:

$$\hat{q}(t) = e^{-\beta t} \qquad (7.3)$$

We may generalize this framework by assuming that there are n regions to explore say, and each region contains either one deposit or no deposits. Let p_j be the prior probability that region j contains a resource deposit. Then the probability $P(t_j)$ of discovering a resource deposit in region j, if time t_j is spent in exploring that region is

$$P(t_j) = p_j(1 - e^{-\beta_j t_j}) \qquad (7.4)$$

where the model (7.2) is used and it is assumed that exploration rates are constant. The optimal decision problem arises as follows. We have T hours available for exploration. How is one to optimally allocate search effort in exploring the various regions when the search efforts are independent statistically and subject to the probability law (7.4)? From past experience p_j may be estimated or by suitable Bayesian methods, posterior (or revised) estimates may be utilize prior to solving for the optimal decision variables t_j (j=1,2,...,n). The optimization model then becomes

$$\underset{t_j}{\text{Max}} \; J = \sum_{j=1}^{n} p_j (1 - e^{-\beta_j t_j}) \qquad (7.5)$$

subject to

$$\sum_{j=1}^{n} t_j = T, \quad t_j \geq 0$$

One could solve this problem directly by Kuhn-Tucker theory of nonlinear programming, since the objective function is strictly concave in t_j and the restrictions are linear. Hence the optimal solutions t_j^*, λ^* satisfy the following necessary conditions which are also sufficient:

$$t_j^* \le \frac{1}{\beta_j} \ln\left(\frac{\beta_j p_j}{\lambda}\right), \qquad t_j^* \ge 0; \; j=1,2,\ldots,n$$

$$\sum_{j=1}^{n} t_j^* = T, \qquad \lambda^* \ge 0$$

If for any region j, $t_j^* > 0$, then

$$t_j^* = \frac{1}{\beta_j} \ln\left(\frac{\beta_j p_j}{\lambda}\right) \tag{7.6}$$

and if this interior solution (7.6) holds for those $j=1,2,\ldots,n$, with $n_1 \le n$ which is in the complete optimal solution of the search model (7.5) then the optimal value λ^* of the Lagrange multiplier λ may be obtained as

$$\lambda^* = \exp\left\{\frac{A_1 - T}{A_2}\right\}, \qquad \lambda^* > 0$$

where $\quad A_1 = \sum_{j=1}^{n_1} \{[\ln(\beta_j p_j)]/\beta_j\}$

$$A_2 = \sum_{j=1}^{n_1} (1/\beta_j)$$

This model of optimal search can easily be extended to a Cournot-type market game between two players. Danskin [12] and Stone [13] have discussed other game theory applications. In place of the decision variables t_j let us introduce two sets of strategies x_j and y_j for the two players. The x-player maximizes his net payoff $J_1 = \sum_{j=1}^{n} x_j r_j$ subject to the normalizing constraints $\sum_{j=1}^{n} x_j = 1$, $x_j \ge 0$ but the parameter r depends on the strategies pursued by the y-player. Let this dependence be of the form

$$r_j = v_j e^{-\beta_j y_j} \tag{8.1}$$

where v_j, β_j are positive constant parameters. The y-player's objective is to minimize the payoff of the first player since the total market is fixed. Let $F(x,y)$ denote the payoff

$$F(x,y) = \sum_{j=1}^{n} [v_j x_j \exp(-\beta_j y_j)]$$

then we have the following optimization problem

$$\text{Max Min } F(x,y)$$
$$\quad x \quad y$$

subject to

$$\sum_{j=1}^{n} x_j = 1, \quad x_j \geq 0 \tag{8.2}$$

$$\sum_{j=1}^{n} y_j = 1, \quad y_j \geq 0$$

Since the objective function $F(x,y)$ is concave in the strategies of the maximizing player and convex in that of the minimizing player, we know that a pair (x^0, y^0) of pure strategy vectors exists such that the following saddle point inequalities hold

$$F(x^0, y) \geq F(x^0, y^0), \text{ for all feasible vectors } y$$

and

$$F(x, y^0) \leq F(x^0, y^0), \text{ for all feasible vectors } x.$$

Let λ and μ be the two Lagrange multipliers for $\Sigma x_j = 1$ and $\Sigma y_j = 1$, then there exists by Kuhn-Tucker theory a pair of positive numbers λ^0, μ^0 such that the following holds:

$$v_j \exp(-\beta_j y_j^0) \begin{cases} = \lambda^0, & \text{if } x_j^0 > 0 \\ \leq \lambda^0, & \text{if } x_j^0 = 0 \end{cases}$$

and

$$v_j \beta_j x_j^0 \exp(-\beta_j y_j^0) \begin{cases} = \mu^0, & \text{if } y_j^0 > 0 \\ \leq \mu^0, & \text{if } y_j^0 = 0 \end{cases}$$

Two points may be noted. One is that the parameters v_j, β_j may not be known to the two players and in that case some type of learning behavior may have to be introduced using a Bayesian or other suitable adaptive rules. Secondly, the form of the r_j function in (8.1) may be such that the payoff function $F(x,y)$ is not concave in x, given y and also not convex in y, given x. In this case the pure strategy vectors may not exist, i.e., the two reaction curves of the two players may not intersect. We may then have to seek optimal solutions in mixed strategies, where the latter are probability mixtures of pure strategies in the domain defined by the constraints of (8.2). By construction the mixed strategies are risk averse in the sense that they attempt to seek solutions which are in some sense "the best of the worst".

Example 5 (Stability of stochastic decision rules)

Whenever a difference or differential equation involves a stochastic variable X(t), which obeys a stochastic process, there arise two types of stability questions which have important implications for economic policy models. One concerns the mean $E\{X(t)\}$ of the process and the differential/difference equations satisfied by it. Does it have a particular stability characteristic? The second question involves random parameters in the dynamic equations governing the economic systems. This issue is important in the context of those econometric models where the parameters to be estimated are not unknown constants but random themselves. Models of expectation formation, lags in adjustment behavior or conjectural variations among Cournot players are all sources of models with random parametric coefficients [15].

The first type of question arises very naturally in many economic models where nonlinearities are present in the stochastic differential equations and linear approximations are made to analyze the time behavior of the solution when time t gets very large. We seek to determine in this framework in some approximate way the time-dependent moments associated with the solution process of the nonlinear differential equations of the Ito type. We follow Soong [14] to present some examples of this procedure.

Let us consider a vector differential equation system in Ito form

$$dX(t) = F(X(t),t) \, dt + G(X(t),t) \, db(t) \qquad t \geq t_0 \qquad (9.1)$$

where $X(t)$ is an n-dimensional vector process, $F(\cdot)$ is a matrix function of order n by n, $G(\cdot)$ a matrix function of order n by m and $B(t)$ is an n-dimensional vector Wiener process with components $B_j(t)$, $j=1,2,\ldots,m$ having the following properties

$$E\{\Delta B_j(t)\} = E\{B_j(t+\Delta t) - B_j(t)\} = 0$$

$$(9.2)$$

$$E\{\Delta B_i(t) \, \Delta B_j(t)\} = 2D_{ij}\Delta t, \qquad t \geq t_0$$

$$i,j=1,2,\ldots,m$$

CHAPTER II

The transition probability density of X(t) denoted by $f(x,t|x_0,t_0)$ satisfies the Fokker-Planck equation

$$\frac{\partial f(x,t|x_0,t_0)}{\partial t} = - \sum_{j=1}^{n} \frac{\partial}{\partial x_j} [\alpha_j(x,t) f]$$

$$+ \frac{1}{2} \sum_{i,j=1}^{n} \frac{\partial^2 [\alpha_{ij}(x,t)f]}{\partial x_i \partial x_j} \qquad (9.3)$$

with the initial condition

$$f(x,t_0|x_0,t_0) = \prod_{j=1}^{n} \partial(x_j - x_{0j}),$$

where $\partial(x_j - x_{j0}) = 1$ if $x_j = x_{jo}$ but zero otherwise and $\alpha_j(x,t)$, $\alpha_{ij}(x,t)$ are as follows

$$\alpha_j(x,t) = \lim_{\Delta t \to 0} \frac{1}{\Delta t} E\{\Delta X_j(t) | X(t) = x\}$$

$$\alpha_{ij}(x,t) = \lim_{\Delta t \to 0} \frac{1}{\Delta t} E\{\Delta X_i(t) \Delta X_i(t) | X(t) = x\}$$

$$= 2 \sum_{k,s=1}^{m} D_{ks} G_{ik}(x,t) G_{js}(x,t)$$

$$= 2(GDG^T)_{ij}, \qquad i,j=1,2,\ldots,n$$

where D denotes the m by m matrix whose ij-th element is D_{ij} and the superscript T denotes the transpose. The Fokker-Planck equation (9.3) now becomes

$$\frac{\partial f}{\partial t} = - \sum_{j=1}^{n} \frac{\partial}{\partial x_j} [\alpha_j(x,t)f] + \sum_{i,j=1}^{n} \frac{\partial^2}{\partial x_i \partial x_j} [(GDG^T)_{ij} f] \qquad (9.4)$$

As an example let $\alpha_j(x,t) = a_j x_j$, a_j being constant. Then

$$\frac{\partial f}{\partial t} = - \sum_{j=1}^{n} a_j \frac{\partial(x_j f)}{\partial x_j} + \sum_{i,j=1}^{n} D_{ij} \frac{\partial^2 f}{\partial x_i \partial x_j}$$

Then using the characteristic function for the transition probability one can calculate

$$f(x,t|x_0,t_0) = (2\pi)^{-/2} |K|^{-\frac{1}{2}} \exp[-\tfrac{1}{2}(x-m)^T K^{-1}(x-m)]$$

where the components of the mean vector m and covariance matrix K are

$$m_j = x_{j0} \exp(a_j t)$$

$$k_{ij} = -\frac{2D_{ij}}{(a_i + a_j)}[1 - \exp(a_i + a_j)t] \qquad i,j=1,2,\ldots,n.$$

From the Fokker-Planck equation (9.4) one can determine the equation satisfied by any moment of $X_1, X_2, \ldots, X_n$. For example define

$$h(X,t) = X_1^{q_1} X_2^{q_2} \ldots X_n^{q_n}$$

Then the moment

$$m_{q_1 q_2 \ldots q_n}(t) = E\{h(X,t)\}$$

can be determined from the differential equation

CHAPTER II

$$\frac{dE\{h\}}{dt} = \sum_{j=1}^{n} E\{\alpha_j \frac{\partial h}{\partial X_j}\} + \sum_{i,j=1}^{n} E\{(GDG^T)_{ij} \frac{\partial^2 h}{\partial X_i \partial X_j}\}$$

$$+ E\{\frac{\partial h}{\partial t}\} \qquad (9.5)$$

For example if we have a first-order Ito equation

$$dX(t) + [a + W(t)] X(t) = 0$$

$$t \geq 0, \; X(0) = X_0 \text{ given}$$

where $W(t)$ is a Gaussian process with zero mean and covariance $2D_{11} \delta(t-s)$, where $\delta(t-s) = 1$ for $t=s$ and zero otherwise. In this case $n = m = 1$, $\alpha_1 = -aX(t)$ and $(GDG^T)_{11} = -D_{11} X^2(t)$. On using (9.5) we obtain the moment equation

$$\dot{m}_j(t) = \frac{dm_j(t)}{dt} = -j[a - (j-1) D_{11}] \, m_j(t) \qquad (9.6)$$

$$j=1,2,\ldots$$

where $m_j(t) = h(X,t) = E\{X^j(t)\}$. It is clear that the moment equation (9.6) has the general solution

$$m_j(t) = E\{X_0^j\} \cdot \exp\{-jt(a-(j-1))\}$$

for which it is clear that for $a < 1$, $m_1(t)$ tends to zero but $m_2(t) \to \infty$.

One implication of this result is that in the moment space the convergence of the first and second moment as $t \to \infty$ may very well differ from their deterministic counterparts. Thus consider another example of a scalar linear differential equation with additive noise:

$$dX(t) = a_0 X(t) \, dt + a_1 X(t) \, dB(t) \tag{9.7}$$

where $X(0) = X_0$ is given, the coefficients a_0, a_1 are fixed and the components $B(t)$ of $dB(t)$ denotes a Wiener process as in (9.2). The deterministic model corresponding to (9.7) is

$$dX(t)/dt = a_0 X(t)$$

whose solution is not stable if a_0 is positive since $X(t)$ tends to infinity as t tends to infinity. But the stochastic model (9.7) has the general solution

$$X(t) = X_0 \exp\left[(a_0 - \tfrac{1}{2}a_1^2)\, t + a_1 B(t)\right]$$

which is stable in its deterministic part if $0 < a_0 < a_1^2/2$. Also we obtain in the moment space:

$$E\{X^k(t)\} = (X_0^k) \exp\left[(a_0 - \tfrac{1}{2}a_1^2)\, kt + \tfrac{1}{2} a_1^2 k^2 t\right]$$

so that if a_0 is negative the first moment $E\{X(t)\}$ converges to zero but the second moment $E\{X^2(t)\}$ explodes to infinity as $t \to \infty$ unless $-a_0 > a_1^2/2$.

The method of moments is also very useful in approximation of nonlinear renewable resources [2] where the growth of a natural resource like fish is given in deterministic form as a sclar nonlinear differential equation

$$\dot{x} = f(X) - h$$

where h is the rate of harvest and $f(X)$ is usually taken as a logistic function. If a linear decision rule relating h to X is adopted then we obtain

$$dX(t) = g(X)\,dt + dB(t)$$

where $B(t)$ in $dB(t)$ follows a Wiener process and $g(X)$ is a nonlinear function of X. For instance let

$$g(X) = -X(t)(1 + aX^2(t))$$

then by applying the equation (9.5) obtained before we can derive the moment equation

$$\dot{m}_k(t) = -k(m_k + am_{k+2}) + Dk(k-1)\,m_{k-2}$$

where $m_k(t) = m_k = E\{X^k(t)\}$.

Thus we obtain for the first two moments

$$\dot{m}_1(t) = -m_1(t) - am_3(t)$$

$$\dot{m}_2(t) = 2[D - m_2(t) - am_4(t)]$$

It is clear that the third and fourth moments m_3, m_4 have direct impact on the time profile of the first two moments.

A second type of stability question arises when we have a continuous model in the state variable $x(t)$:

$$\dot{x}(t) = Ax(t) + Bu(t) \qquad (10.1)$$

but the control variables $u(t)$ is applied with a lag or delay, i.e.,

$$u(t) = Gx(t-\tau), \quad \tau \neq 0 \qquad (10.2)$$

Here A,B,G are suitable constant matrices and $x(t)$, $u(t)$ denote n-tuple and m-tuple vectors. Reasons for this lag in behavior may be several, e.g., (a) information about $x(t)$ may only be availabe in discrete time, (b) the agents may have perception or reaction lags, (c) as Cournot players the lags may reflect caution in adjusting to new information. By combining (10.2) with (10.1) we obtain for $C = BG$:

$$\dot{x}(t) = Ax(t) + Cx(t-\tau) \qquad (10.3)$$

Suppose with zero control $u(t) = 0$, the system (10.1) is stable in the sense that all the characteristic roots of A have negative real parts. Under what restrictions on the elements of G and the delay τ the controlled system would still be asymptotically stable, i.e., $x(t)$ tends to a finite constant as $t \to \infty$? This question is important for economic policy models, since the application of government expenditure policy with a lag may otherwise be destabilizing, as was emphasized by Phillips [4] in a macrodynamic Keynesian model.

Consider a scalar example of (10.3) as:

$$\dot{x}(t) = -ax(t) - cx(t-\tau) \qquad (10.4)$$

The characteristic quasi-polynomial equation is:

$$\phi(\lambda) = \lambda + a + ce^{-\lambda\tau} = 0 \qquad (10.5)$$

If c is zero and a is positive, then clearly $x(t)$ is stable. But for nonzero c and τ, the various regions in the solution space are characterized by the real and complex values of λ in (10.5). Now let the quasi-polynomial $\phi(\lambda)$ have a purely imaginary root iq with $i = \sqrt{-1}$

$$iq + a + ce^{-iq\tau} = 0$$

or,

$$iq + a + c(\cos \tau q - i \sin \tau q) = 0$$

On separating the real and imaginary parts and solving for a and c we get

$$c = q \sin \tau q, \quad a = (-q \cos \tau q)(\sin \tau q)^{-1}$$

where

$$D: \{a + c \cos \tau q = 0, \quad q - c \sin \tau q = 0\} \qquad (10.6)$$

defines a curve on which all the roots λ of the quasi-polynomial (10.5) are purely imaginary. If $c\tau$ exceeds unity we get $\cos \tau q < 0$ on the curve defined by D in (10.6) and hence we get unstable oscillations. Also when $c\tau > 0$ we have at least one root with a positive real part and then $x(t)$ tends to be explosive.

It is clear that if the variations in the parameter c are partly stochastic, the regions of the solution space defined by the quasi-polynomial in (10.5) have to be carefully identified in order to secure asymptotic stability in the mean square sense [15,16].

References

1. Tintner, G. and J.K. Sentupta. Stochastic Economics: Stochastic Processes, Control and Programming. New York: Academic Press, 1972.
2. Mangel, M. Decision and Control in Uncertain Resource Systems. New York: Academic Press, 1985.
3. Malliaris, A.G. and W.A. Brock. Stochastic Methods in Economics and Finance. Amsterdam: North Holland, 1982.
4. Phillips, A.W. Stabilization Policy in a Closed Economy. Economic Journal 64 (1954), 290-323.
5. Sengupta, J.K. Optimal Decisions Under Uncertainty: Methods Models and Management. New York: Springer Verlag, 1985.
6. Parlar, M. Optimal Dynamic Service Rate Control in Time Dependent M/M/S/N Queues. International Journal of Systems Science 15 (1984), 107-118.
7. Goodwin, G.C. and R.L. Payne. Dynamic System Identification: Experiment Design and Data Analysis. New York: Academic Press, 1977.
8. Kailath, T. Some Alternatives in Recursive Estimation. International Journal of Control 32 (1980), 311-328.
9. Lewis, T.R. Attitudes Towards Risk and the Optimal Exploitation of an Exhaustible Resource. Journal of Environmental Economics and Management 4 (1977), 111-119.
10. Andersen, P. Commercial Fisheries under Price Uncertainty. Journal of Environmental Economics and Management 9 (1982), 11-28.
11. Pindyck, R.S. Uncertainty in the Theory of Renewable Resource Markets. Review of Economic Studies 51 (1984), 289-303.
12. Danskin, J.M. The Theory of Max-Min and its Application to Weapons Allocation Problems. New York: Springer-Verlag, 1967.

13. Stone, L.D. Theory of Optimal Search. New York: Academic Press, 1975.

14. Soong, T.T. Random Differential Equations in Science and Engineering. New York: Academic Press, 1973.

15. Mori, T. and E. Noldus. Stability Criteria for Linear Differential-Difference Systems. International Journal of Systems Science 15 (1984), 87-94.

16. Sengupta, J.K. Information and Efficiency in Economic Decision. Boston: Martinus Nijhoff Publishers, 1985.

CHAPTER III

Recent Economic Models in Applied Optimal Control

1. Introduction

Recent economic applications of optimal and adaptive control theory have increased our understanding of the meaning of optimality in a dynamic horizon e.g., the shadow prices or adjoint variables along the optimal trajectory, their stability properties and the implications of a steady state value. Optimal control models are now increasingly applied to newer areas of economic theory, of which some illustrative examples would be discussed here as follows:

   A. Models of differential market games,

   B. Stabilization policy in cartelized markets,

   C. Adaptive controls in economic models and,

   D. Models of renewable resources under uncertainty.

2. Conjectural Equilibria in Cournot-Nash Games

The framework of differential games comes very naturally when one needs to model a market where there exist two or more rivals, each with output, price or mixed strategies. One of the earliest market models in this framework is the limit pricing theory, where a monopolist is concerned with the possibility that a second firm may enter the market. Since the perception of the potential entrant concerning the market and the cost-structure of the existing firm is crucial to the dynamic entry equation, one has to analyze the role of uncertainty and limited information to see how they affect the equilibrium solution of the differential game model. Some possible answers have been attempted in the current literature. The first attempt due to Milgrom and Roberts [1] analyses the various expectations of the potential entrant concerning the price cost behavior of the monopolist. As Kreps and Spence [4] have pointed out, this may lead to equilibrium solutions which are very sensitive to the initial conditions and perceptions of the potential entrant. A second approach by Dixit [2] shows how changing expectations of the post-entry market competition may severly affect the prospect of potential entry e.g., entry deterrence may obtain with a high probability if the entrant rightly or wrongly perceives that post-entry competition will be very severe in terms of prices, subject only to the limitations imposed by output capacities. A third approach by Sengupta [3] provides a stochastic view of entry, which shows the asymmetrical influence of risk aversion on the monopolist, which by reducing the short-run profit-maximizing price may act as entry-deterrence.

Although the limit pricing model can be viewed as a dynamic variant of Cournot-Nash equilibria, the two crucial postulates of Cournot markets are not specifically analyzed here e.g., conjectural variations and the perceived influence of each player on the market. Yet it is clear that in a stochastic environment of future demand and potential entry, these two factors may play a crucial role. We consider a few formulations in this regard.

A dynamic formulation of conjectural equilibrium which leads in its steady state to a Cournot solution is due to Kamien and Schwartz [5] and Fershtman and Kamien [6]. They consider a dynamic market as

$$\dot{x} = f(x,u,t), \quad x(0) = x_0 \qquad (1.1)$$

with x and u as the state and control vectors and a dot denotes the time derivative. Each player chooses a component u_i of the control vector to maximize his payoff function :

$$\max_{u_i} J_i = \int_0^T F_i(x,u,t) dt \qquad i = 1,2,\ldots,n \qquad (1.2)$$

Appropriate conditions of differentiability, boundedness and concavity of the functions $f(\cdot)$ and $F_i(\cdot)$ are assumed to assure that the necessary conditions are sufficient. Depending on the information structure available to each player, three types of strategy choices are used:

$$u_i = g_i(x_0, x, t): \text{ closed-loop no memory policy,}$$
$$u_i = g_i(x, t): \text{ feedback policy,} \qquad (1.3)$$
$$u_i = g_i(x_0, t): \text{ open-loop policy.}$$

As the functional form of $g_i(\cdot)$ indicates that the feedback policy when it exists as an optimal strategy, has two most desirable properties: optimal control depends essentially on the current state variables and hence it is rather easy to update with additional information. The open loop policy on the otherhand does not necessarily possess the feedback property along the optimal trajectory except e.g., when the control model is of the LQG (linear quadratic Guassian) form [7]. We define a Nash equilibrium in feedback strategies as an n-type of decision rules $(g_1^*(x,t), g_2^*(x,t), \ldots, g_n^*(x,t))$ such that the following inequality holds for every (x_0, t_0):

$$J_i(g_1^*, \ldots, g_n^*) \geq J_i(g_1^*, \ldots, g_{i-1}^*, g_i, g_{i+1}^*, \ldots, g_n^*) \qquad (1.4)$$

for all $i = 1, 2, \ldots, n$

In other words, g_i^* specifies the best response of player i ($i=1,2,\ldots,n$) to the strategies $(g_1^*, \ldots, g_{i-1}^*, \ldots, g_n^*)$ of the other n-1 players. Now we introduce the functions $\bar{h}_i(x,t) = (h_1(x,t), \ldots, h_{i-1}(x,t), h_{i+1}(x,t), \ldots, h_n(x,t))$ as the conjectures of firm i about the behavior of its rivals and assume that it satisfies some regularity conditions known as Lipschitz conditions (i.e., continuity and boundedness). Then player i solves the optimal decision problem as:

$$\max_{u_i} J_i = \int_0^T F_i(x, \bar{h}_i(x,t), u_i) dt$$

subject to

$$\dot{x} = f(x, \bar{h}_i(x,t), u_i) \qquad (1.5)$$

$$x(0) = x_0$$

Let $\{\tilde{u}_i(t) | 0 \leq t \leq T\}$ be the optimal solution trajectory, when it exists. Given the $(n-1)$ conjecture functions $\bar{h}_i(x,t)$ and the optimal control path $\tilde{u}_i(t)$, one could compute the optimal path $\tilde{x}(t)$, $0 \leq t \leq T$ of the state vector on the basis of the dynamic equation of motion in (1.5). Thus if the time path of the state variable $\tilde{x}(t)$ is available (or expected), then the player i can easily construct the expected time path of control $(\tilde{u}_1(t), \tilde{u}_2(t), \ldots, \tilde{u}_n(t))$ which may be denoted by

$$R_i(h) = R_i(h_1, h_2, \ldots, h_n) = R_i(\bar{h}_1, \bar{h}_2, \ldots, \bar{h}_n)$$

or, $\qquad (1.6)$

$$R(h) = (R_1(h), R_2(h), \ldots, R_n(h))$$

In other words we can define for every n-tuple of conjectures $h = (h_1, h_2, \ldots, h_n)$ a vector $R(h)$ such that $R_j(h) = R_j(\bar{h}_j)$. Note that this derivation (1.6) implies two types of consistency requirements. One is that the conjectures of any two distinct players k and j about the behavior of firm i are identical. The second is that only under feedback optimal strategies that the expected time path of the state variables $\tilde{x}(t)$ may generate responses that are equilibriating.

Given the optimal control model (1.5) for each Cournot player and the consistency requirement defined by (1.6), Fershtman and Kamien [6] define two types of equilibrium: a conjectural equilibrium (CD) and a perfect conjectural equilibrium (PCE). A CE is an n-tuple of conjectures $h^* = (h_1^*, h_2^*, \ldots, h_n^*)$ such that $R_i(h^*) = R_j(h^*)$ for every $i \neq j$ $(i,j = 1,2,\ldots,n)$. A PCE is an equilibrium solution if CE holds for all possible initial values (x_0, t_0). They then show that every conjectural equilibrium $(h_1^*(x,t), \ldots, h_n^*(x,t))$ constitutes a closed loop no memory Nash equilibrium and vice versa. This implies that every conjectural equilibrium h^* satisfies the inequality

$$J_i(h^*) \geq J_i(h_1^*, \ldots, h_{i-1}^*, g_i, h_{i+1}^*, \ldots, h_n^*) \tag{1.7}$$

for all $i=1,2,\ldots,n$

for every possible closed-loop no-memory strategy g_i but since the starting point x_0 may vary, the conjectural equilibrium $h^* = (h_1^*, h_2^*, \ldots, h_n^*)$ is a PCE if and only if it is a Nash equilibrium in feedback strategies.

As an example they consider a duopoly market for a single good where $u_i(t)$ is the output of player $i=1,2$ and the dynamics of market price $p(t)$ is specified by the linear differential equation model:

$$\dot{p} = s(a-b(u_1+u_2)-p)$$
$$p(0) = p_0; \quad 0 < s < \infty \tag{1.8}$$

cost functions are assumed quadratic i.e., $C(u_i) = cu_i + \tfrac{1}{2} u_i^2$. The payoff functions are

$$\max_{u_i} J_i = \int_0^\infty [\exp(-rt) \cdot \{pu_i - cu_i - \tfrac{1}{2}u_i^2\}] dt \tag{1.9}$$

where r is a positive discount factor assumed exogenous and s denotes the speed of price adjustment. Since this is in the format of an LQG model of optimal control, the optimal feedback strategies $u_i^*(t)$ exist for each player i and are given by

$$u_i^*(t) = \{1-bsk(t)\}p(t) + bsm(t) - c \qquad (1.10)$$

where

$$k(t) = (6s^2b^2)^{-1}[r+4bs+2s-((r+4bs+2s)^2 - 12s^2b^2)^{\frac{1}{2}}]$$

$$m(t) = (r-3b^2s^2k(t)+s+2bs)^{-1}[-ask(t)+c-2bsck(t)]$$

It is clear that the PCE price path exists and converges to a unique steady state price p^* where

$$p^* = [2b(1-bsk(t))+1]^{-1}[a+2b(c-bsm(t))], \qquad (1.11)$$

For more specific results one has to assume some form of dependence of the conjectures e.g., if we assume $\partial u_i / \partial u_j = \alpha$, $(i \neq j)$, where α is some positive constant, the solutions $u_i^*(t)$ and p^* would be somewhat different.

While it is clear that the concept of a dynamic conjectural equilibrium is fundamental to any generalization of the Cournot model of a differential market game, it is by no means apparent [8] that the requirements of consistency given by (1.6) would hold in many reasonable cases. For one thing, the conjectures and the process of their formation are not related to any subjective probability structue or their appropriate revision. From the viewpoint of information requirement, the specification of optimal (or suboptimal) strategies under limited information must incorporate some form of uncertainty. Secondly, even when CE conjectures $h^* = (h_1^*, \ldots, h_n^*)$ fail to exist, mixed strategy solutions may very well be introduced by the tracing procedure approach of Harsanyi [9], where each player instead of treating his opponent's most recently calculated best play as a 'new probability distribution', only uses it to modify an estimate of his opponent's strategy choices. Under certain conditions, these estimates are shown by Harsanyi to converge to a unique equilibrium point of the Cournot game, provided the latter exists as a mixed strategy vector. Thirdly, the optimal feedback solutions which are known to exist for the simple example of the duopoly model given by (1.8) through (1.11) may fail to hold even in very simple cases. Consider for instance a static case with zero costs of production and a demand function

$$p = b(u_1+u_2)^{-1}; \quad b>0$$

with profits $\pi_i = pu_i$. The two optimal reaction functions of the two players are

$$u_1 = u_1 + \bar{u}_2, \quad u_2 = u_2 + \bar{u}_1 \tag{1.12}$$

where $\bar{u}_2$ is the conjecture of player one about his opponent's strategy and $\bar{u}_1$ that of player two. It is clear that there is no Cournot equilibrium solution other than $u_1 = 0 = u_2$. Following a Nash equilibrium however one could still define a reasonable solution by maximizing the product $(\pi_1\pi_2)$ where $\pi_2 = b-\pi_1$. Clearly the optimal solution is $\pi_1^* = b/2 = \pi_2^*$ with $u_1^* = u_2^* > 0$. Note that a dynamic analogue of this static model can be formulated as

$$\max J_i = \int_0^T [\exp(rt) \cdot pu_i] dt$$

subject to

$$\dot{p} = s[(u_1+u_2)^{-1}b-p]$$
$$p(0) = p_0$$

CHAPTER III

where (1.12) may be interpreted as a steady state approximation. It is clear that a CE price path converging to a unique steady state Cournot equilibrium price cannot be specified here.

There is a more fundamental issue with the concept of conjectural equilibrium, when the interdependence of strategies through such terms $\partial u_i / \partial u_j$ ($i \neq j$) is implicitly known or guessed by the players. In such cases the players may seek implicit cooperation by "subjective random devices" as proposed by Aumann [10] or, by "correlated strategies" as proposed by Moulin [11].

The Cournot-Nash solution of a noncooperative game can be improved in many cases by recognizing that the players are interdependent. This is so because the Cournot equilibrium solution has two major features which favor implicit cooperation. First, although no single firm (i.e. player) at the equilibrium point has any incentive to move out, there exist general moves by which all firms can increase their profits simultaneously. Since collusive agreements are not permitted in the Cournot model, questions arise whether "subjective random devices" (Aumann 1973), or "public lotteries" [11] can be designed to improve profits for all firms. Second, the reaction function theory of reaching a Cournot-Nash (CN) equilibrium solution in a two-person game explicitly allows an adaptive trial and error process before reaching the equilibrium, where one player takes the strategy of the other player as a parameter. However this parameter has to be estimated (or forecast) each time one player makes his move. Such estimates or predictions can in many cases help them realize that they can improve the noncooperative solution and these may be formed either through the "public" information available to both players or through "private" information subjectively held by each player. Correlation of strategies here, introduced through such information channels, may again improve profits for both firms.

Whereas the first aspect of interdependence has been emphasized in the recent research work on correlated strategies in a CN framework, the second has received very little attention. Our objective here is to analyze this second aspect of interdependence in a two-person CN game and to show how correlation in a statistical sense may improve the solution. Our view of correlation is different from those defined by Aumann [10] and Moulin [11] in two respects. One is that our correlation is identical with the standard statistical concept of the coefficient of correlation, which lies between zero and one in absolute value. Depending on the probability distributions of strategy choices, this correlation may therefore be statstically tested against realized outcomes. This is most convenient, since for nonparametric cases the analogous concept of rank correlation coefficient may be directly utilized, thus making it suitable for qualitative games. The second aspect of our correlation concept is the fact that it leads very naturally to the problem of maximizing or minimizing a ratio which is known as the Rayleigh quotient in applied mechanics. Since the Rayleigh quotient optimization is usually done by the Rayleigh-Ritz algorithm for approximate calculation of the minimum or maximum eigenvalues, our framework of correlated strategies may be viewed in suitable cases as an iterative process of successive improvement. If the successive iterations are terminated or truncated at a certain stage due to additional restrictions on the strategy space, their implicit costs can be directly evaluated. This consideration is important for the class of constrained games under partially controllable strategies [12]. Let us start with some examples of correlation in duopoly games.

The first example is from Aumann [10], where correlation is introduced through randomized strategies, where a randomized strategy is viewed as a random variable with values in the pure strategy space, rather than as a

CHAPTER III

distribution over pure strategies. Consider the following two-person nonzero-sum game with the payoff matrix

$$\begin{bmatrix} 2,1 & 0,0 \\ 0,0 & 1,2 \end{bmatrix}$$

There are exactly three CN equilibrium points, e.g. two in pure strategies yielding (2,1) and (1,2) respectively and one in mixed strategies, yielding (2/3, 2/3). The payoff vector (3/2, 3/2) is not achievable at all if the mixed strategies are objectively determined. It is however achievable in correlated strategies as follows: one fair coin is tossed, players I and II play top and left respectively; otherwise they play bottom and right. The higher payoff vector (3/2, 3/2) is thus achievable by correlation and it is also an equilibrium point, since neither player can gain by a unilateral change. It is clear that the problem can be generalized to an n-person game with more than two strategies for each player.

Consider a second example of a market game with two duopolists each supplying outputs x and y with respective profit functions

$$u = a_1 x - bx^2 - bxy$$
$$u = a_2 x - by^2 - bxy$$

where total revenue is rx and ry respectively and total costs are $c_1 x$ and $c_2 y$, with c_1, c_2 being fixed positive constants and $r = a-b(x+y)$, $b > 0$ indicates the demand function relating price and quantity; also $a_i = a-c_i$ (i = 1,2) are positive parameters. The standard CN equilibrium with no correlation of strategies is obtained from the solution of the two linear reaction functions

$$2bx = a_1 - b\hat{y}$$
$$2by = a_2 - b\hat{x}$$
(2.1)

where $\hat{y}$ is the first player's guess of the second player's output strategy and $\hat{x}$ is the second player's guess of the first player's strategy. Now assume that the guesses or estimates $\hat{y}$, $\hat{x}$ are drawn from a bivariate normal density $f(x,y)$ with means μ_x, μ_y, variances σ_x^2, σ_y^2 and the correlation coefficient ρ, where ρ is positive. Player I then maximizes the conditional expected profit

$$E(u|x) = a_1 x - bx^2 - bxE(y|x) \qquad (2.2)$$

and player II does likewise by maximizing

$$E(v|y) = a_2 y - by^2 - byE(x|y) \qquad (2.3)$$

This leads to the following reaction functions:

$$2bx(1 + \rho\sigma_y/\sigma_x) = a_1 + \frac{b\rho\sigma_y \mu_x}{\sigma_x} - b\mu_y$$

$$2by(1 + \rho\sigma_x/\sigma_y) = a_2 + \frac{b\rho\sigma_x \mu_y}{\sigma_y} - b\mu_x$$
(2.4)

Note that if the correlation coefficient is zero and the mean parameters μ_y, μ_x are replaced by their estimates $\hat{y}$ by player I and $\hat{x}$ by player II, then the system (2.4) reduces to the standard CN equilibrium defined in (2.1). Now

CHAPTER III

assume without loss of generality that $\sigma_x^2 = \sigma_y^2 = 1.0$ and that ρ is not zero. Then we obtain

$$\left.\frac{\partial^2 E(u|x)}{\partial x^2}\right|_{\substack{x=\hat{x} \\ y=\mu_y=\hat{y}}} = -2b(1+\rho) = \left.\frac{\partial^2 E(v|y)}{\partial y^2}\right|_{\substack{y=\hat{y} \\ x=\mu_x=\hat{x}}}$$

$$\left.\frac{\partial^2 E(u|x)}{\partial y \partial x}\right|_{\substack{x=\hat{x} \\ y=\mu_y=\hat{y}}} = -b = \left.\frac{\partial^2 E(v|y)}{\partial x \partial y}\right|_{\substack{y=\hat{y} \\ x=\mu_x=\hat{x}}}$$

Hence the Hessian matrix at $x = \hat{x}$, $y = \hat{y}$ becomes

$$H = \begin{bmatrix} -2b(1+\rho) & -b \\ -b & -2b(1+\rho) \end{bmatrix}$$

which is negative definite if

$$(1+\rho)^2 > \tfrac{1}{4}$$

For any nonnegative ρ with $0 \leq \rho \leq 1$ this is always fulfilled but if ρ is sometimes negative, then we must have

$$\rho > -\tfrac{1}{2} \qquad (2.5)$$

Thus if the correlation coefficient ρ is negative and fails to satisfy the condition (2.5), we have the situation that the CN equilibrium solution (x =

$\hat{x}$, $y = \hat{y}$) can be improved. This result has several interesting interpretations. First, both the players would have positive incentives to provide tacit signalling mechanisms or "subjective random devices" by which the probability of the event ($\rho < -\frac{1}{2}$) may be made very low or negligible. Since the event ($\rho < -\frac{1}{2}$) is undesirable, this amounts to requiring that the probability of the undesirable event (which is comparable to the probability of ruin) be minimized. Second, in default of not knowing ρ exactly, the best they could do is perhaps to maximize it. One rationale for this is that they may both hope to gain by such implicit coordination in behavior. A second rationale is that under consumers' perceptions the strategies of the two players may appear very similar when correlated.

Consider a third example where the profit function (i.e. payoff) of each player is of the form $\pi = xy$, where x is the strategy of player I and y that of player II. These strategies are however functions of the outcome of a random move, which may be described for example by two uniformly distributed random variables. Restrepo [13] has considered a general class of games with a random move for each player, where the random move is described by two uniformly distributed random variables with values x and y in the unit interval $[0,1]$. Here the strategies are designed such that, having received some information about the outcome of the random move, each player may select one of several courses of actions from a finite set of alternatives. Two interpretations of this type of random strategies are useful in economic models. One is the presence of imperfect knowledge or ignorance about the form of interdependence of the two players i.e. ignorance about each other's reaction functions and the second is the uncertainty of the environment to which both players are subject. Thus there exists a stochastic relation of the form

$$h(\tilde{x}, \tilde{y}) = 0 \qquad (3.1)$$

where the strategies $\tilde{x}$, $\tilde{y}$ are random quantities drawn from some bivariate distribution and the function $h(\cdot)$ is not separable i.e. none of the two players knows if $\tilde{x} = \tilde{x}(\tilde{y})$ is a function of $\tilde{y}$, or $\tilde{y} = \tilde{y}(\tilde{x})$ is a function of $\tilde{x}$. In default of this knowledge each player has to specify optimal strategies in the situation where profits π are random. Following the standard economic theory of risk we assume that each player is risk averse and maximizes an expected utility function of the form

$$z = E[U(\pi)] = E\pi - \frac{\lambda}{2}V(\pi) \qquad (3.2)$$

where E and V denote expectation and variance and λ is a positive parameter related to risk aversion. Let the random variables $\tilde{x}$, $\tilde{y}$ be drawn from a bivariate distribution with means μ_i and variances σ_i^2 (i = 1,2) and correlation ρ. Without any loss of generality it may be assumed that $\sigma_i^2 = 1$. Thus the random moves may be specified as

$$\tilde{x} = \mu_1 + e_1; \; \tilde{y} = \mu_2 + e_2$$

where e_1, e_2 have zero means and a correlation ρ. The players can now choose their optimal strategies in two different ways. One is the standard Cournot approach, where the rival is assumed to be passive and the second is to allow the correlation of their strategies through tacit signals. In the first case the Hessian matrix is

$$H = \begin{bmatrix} -\lambda & 0 \\ 0 & -\lambda \end{bmatrix}$$

which is always negative definite, since $\lambda > 0$. In the second case the Hessian matrix evaluated at μ_1, μ_2 is different i.e.

$$H = \begin{bmatrix} -\lambda & 1-\lambda\rho \\ 1-\lambda\rho & -\lambda \end{bmatrix}$$

It is negative definite only for some special cases depending on the value of λ i.e, if any of the following four conditions hold:

(a) for $0 < \rho < 1$, $\rho > (1/\lambda) - 1$
(b) for $\rho < 0$, $0 < |\rho| < 1 - (1/\lambda)$
(c) for $\rho = 0$, $\lambda > 1$
(d) for $\rho = 1$, $\lambda > \frac{1}{2}$

(3.3)

It is clear that there are many cases when the conditions (3.3) would not hold. First of all, if the risk aversion parameter λ is positive but less than 0.5, the last three cases of (3.3) fail to hold. There is no incentive for joint maximization in such situations. Second, λ must satisfy the condition $0.50 < \lambda < 1$ for case (a), when the correlation is positive but less than one. Thus two things are clear. One is the case when the players are highly risk averse in the sense $\lambda > 0.50$ and the other when they have low risk aversion (i.e. $\lambda < 0.50$). In the first case, positively correlated strategies are helpful for both players, provided $\rho > (1/\lambda) - 1$. If initially ρ is less than the level $k = (1/\lambda) - 1$, it pays both players to raise their level of correlation above k. Improving correlation is a Pareto optimal policy. In the second case when they have low risk aversion (i.e. $\lambda < 0.50$), it is impossible for both players to gain for any value of ρ where $0 \leq |\rho| \leq 1$.

Here raising the level of correlation is not helpful for both, since it would not make H negative definite. We arrive therefore at the following result which appears to be intuitively reasonable: If the players are highly risk averse in the sense $\lambda > 0.50$, they will both have incentive to cooperate and to raise their correlation above the level k, where $k = (1/\lambda) - 1$. But if they are not highly risk averse, they would have no such incentive to correlate their strategies. The first case is typical in many bargaining situations, where the players are more willing to cooperate through a mediator, the second may need compulsory arbitration.

We consider now the case of mixed strategies for both players. Let p and q be column vectors corresponding to x and y before with dimensions m and n respectively. The objective functions of the two players are

$$\begin{aligned} \text{I:} \quad & \max w = p'\theta \\ & p \geq 0 \\ \text{II:} \quad & \max w = z = q'\phi \\ & q \geq 0 \end{aligned} \qquad (4.1)$$

where prime denotes transpose and both θ and ϕ contain elements of payoff not completely known to each player e.g. θ may be influenced in part by q, which is not known to player I who chooses p. Both players have to choose their optimal mixed strategy vectors p and q, when they have ignorance or incomplete information about the vectors θ and ϕ in their objective functions. Two types of incomplete information are analyzed here. The first and the most important case arises when the vectors θ, ϕ are known to be jointly distributed and both players intend to maximize the correlation of their respective payoffs, $R = R(w,z)$. Two motivations may be given for such a correlated strategy by both players. On the one hand, it emphasizes a rational forecasting principle for each player when it has to forecast his opponent's profit levels. A second motivation is that each player assumes the other party to have additional information that he does not have and by maxmizing the correlation $R(w,z)$ of their payoffs he is in effect economizing his cost of search.

A second case of using incomplete information arises when each player seeks to predict so to say the respective random vectors θ, ϕ by their respective regression models:

$$\begin{aligned} \theta &= E\theta(p,q|\hat{q}) + e_1 \\ \phi &= E\phi(p,q|\hat{p}) + e_2 \end{aligned} \qquad (4.2)$$

where e_1, e_2 are zero-mean random vectors with no correlation and unit variances. This situation is most realistic, when each player has observed several past rounds of the game and intends for instance to forecast the parameters θ (for player one) and ϕ (for player two).

For the first case we assume for the sake of convenience that the vectors θ and ϕ are jointly normally distributed $N(\mu,V)$ with mean μ and variance-covariance matrix V where

$$\mu = \begin{pmatrix} \mu_1 \\ \mu_2 \end{pmatrix}, \quad V = \begin{bmatrix} V_{11} & V_{12} \\ V_{21} & V_{22} \end{bmatrix} \qquad (4.3)$$

$\mu_1 = E\theta$, $\mu_2 = E\phi$ and $V = \text{cov}(\theta,\phi)$ is the covariance matrix assumed to be positive definite. Since the players have incomplete knowledge about the random variations for θ and ϕ, they may find it rational to maximize the correlation $R = R(w,z)$ of their payoffs:

$$\underset{p,q \geq 0}{\text{Max}} \quad R = \frac{p'V_{12}q}{[(p'V_{11}p)(q'V_{22}q)]^{\frac{1}{2}}} \tag{4.4}$$

In this case each player implicitly assumes that the correlation R is positive and by maximizing it each would increase his own payoff. For many new markets which are growing in demand over time, this assumption may be most reasonable. The problem of jointly maximizing the correlation R has however the following aspect of indeterminacy: if (p^0, q^0) is any solution of (4.4), so would cp^0, dq^0 for arbitrary constants c and d. To avoid this scale problem it is assumed that

$$p'V_1 p = 1 = q'V_{22}q \tag{4.5}$$

This may be viewed either as a normalization condition on p and q, just like $\Sigma p_i = 1 = \Sigma q_j$, or as a constraint on the distance. In the latter case we may define $p = \tilde{p} - p_0$, $q = \tilde{q} - q_0$, where p_0, q_0 are the initial preassigned estimates and $\tilde{p}$, $\tilde{q}$ are the respective perturbations [12]. Now assume that the nonnegative vectors p* and q* solves the optimization problem (4.4) through (4.5), then it can be easily shown that they satsify the following equations

$$\begin{bmatrix} -\lambda V_{11} & V_{12} \\ V_{21} & -\lambda V_{22} \end{bmatrix} \begin{pmatrix} p^* \\ q^* \end{pmatrix} = 0 \tag{5.1}$$

where $\frac{1}{2}\lambda$ is the common Lagrange multiplier associated with the constraint (4.5). For nontrivial solutions of the homogeneous system (5.1), the determinant of its coefficient matrix must be zero i.e.

$$\begin{vmatrix} -\lambda V_{11} & V_{12} \\ V_{21} & -\lambda V_{22} \end{vmatrix} = 0 \qquad (5.2)$$

If the roots of this determinantal equation are ordered as $\lambda_{(1)} \geq \lambda_{(2)} \geq \cdots \geq \lambda_{(m)}$ where m is the order ($m \leq n$) of the square matrix V_{11}, then we have to choose the maximal root $\lambda_{(1)}$ in the equations (5.1) for p^*, q^* in order to maximize the correlation R. On expanding the determinant in (5.2) one obtains

$$(-1)^m \lambda^{n-m} \left| V_{22} \right| \cdot \left| \lambda^2 V_{11} - V_{12} V_{22}^{-1} V_{21} \right| = 0$$

This shows that (5.2) has at least (n - m) zero roots ($m \leq n$) and its nonzero roots satisfy the polynomial equation:

$$\left| \lambda^2 V_{11} - V_{12} V_{22}^{-1} V_{21} \right| = 0$$

Let R(i) denote the correlation associated with the nonzero characteristic root $\lambda_{(i)}$, $1 \leq i \leq m$, then it is clear that both players' realized payoffs would depend on the actual level of correlation or cooperation R(i) achieved. Starting from the level $\lambda_{(m)}$ and the associated correlation $R(m) = R(\lambda_{(m)})$, the players may improve their profitability position by increasing their correlation till the maximal level R(1) and the associated vectors (p^*, q^*) are reached.

Next consider the conditional regression model (4.2) where the two objective functions are written in their reduced form as player I: $\underset{p'p=1}{\text{Max}} J_1 = p'E(\theta|\hat{q})$

player II: $\underset{q'q=1}{\text{Max}} J_2 = q'E(\theta|\hat{p})$ (6.1)

where

$$E(\theta|\hat{q}) = A\hat{q}, \quad E(\phi|\hat{p}) = B\hat{b}$$ (6.2)

Note that if θ and q are jointly normally distributed with a covariance matrix $S_{\theta q}$, then the conditional regression of θ given $\hat{q}$ is

$$E(\theta|\hat{q}) = \mu_\theta + S_{\theta q} S_{qq}^{-1} (\hat{q}-\mu_q)$$ (6.3)

Setting the mean levels μ_θ, μ_q to zero and S_q to an identity matrix for simplicity, we obtain $A = S_{\theta q}$ in (6.2) and similarly $B = S_{\phi p}$. Clearly, $S_{\theta q}$ and $S_{\phi p}$ represent the correlation of the two players' strategies in their respective subjective estimates. Denoting the Lagrange multipliers by $k/2$ and $r/2$ respectively, one could easily derive the CN equilibrium.

Theorem 1

The CN equilibrium solution at $p = \hat{p}$, $q = \hat{q}$ exists for the problem (6.1) through (6.3) for any fixed weight coefficient ω, $0 < \omega < 1$. It satisfies the complementary eigenvalue problems:

$$(\omega(1-\omega) AB - sI) p = 0$$

$$(\omega(1-\omega) BA - sI) q = 0$$ (6.4)

where $s = kr$ is the eigenvalue. If $m < n$ and the matrix AB is positive definite, then the m eigenvalues are positive and can be written as $s_1 > s_2 \geq \ldots \geq s_m > 0$ with their associated nonzero eigenvectors denoted by (p_i, q_i), $i = 1, 2, \ldots, m$. If the players do not choose the maximum eigenvalue s_1 and the associated vectors (p_1, q_1), then their CN solution can be improved by correlation.

Proof.

From the two reaction functions derived from (6.1), one can set $p = \hat{p}$, $q = \hat{q}$ and obtain the eigenvalue problems. Due to the constraints $p'p = 1 = q'q$, there must exist vectors p and q satisfying the first order condition for each maximization problem. If AB is positive definite, the eigenvalues are all real and positive. It follows that at the maximum eigenvalue $s = s_1$, the players reach their highest conditional payoff. So long as the highest eigenvalue is not achieved, the iterative process will help improve their solution.

Corollary 1.1

If the matrix AB has all poisitve elements and is indecomposable, then by the Perron-Frobenius theorem, the maximum eigenvalue s_1 is positive and the optimal strategy vectors p_1, q_1 have all positive elements.

Corollary 1.2

The minimum eigenvalue s_m and the associated vectors p_m, q_m may be interpreted as a conflict payoff, so that the positive difference $(s_1 - s_m)$ may provide the incentive to cooperate by correlation, if it exceeds the extra computing costs. The propensity to concede in a bargaining process may therefore be viewed as directly proportional to $(s_1 - s_m)$.

CHAPTER III

We have not so far analyzed the dynamic conjectural equilibrium illustrated by the duopoly example (1.8) and (1.9) in its stochastic aspects, when the two outputs are themselves random. A natural example occurs in ecology which deal with stochastic models of growth of competitive and predatory populations which interact with one another. Instead of output we have the size of population of each species and competition for food or fixed habitat replaces the competition in the market place. On the analogy of the birth and death type Markov process [14] applied to study the stochastic interaction between two species in ecology, we may formulate a similar model for a stochastic duopoly market. Let $X(t)$ and $Y(t)$ be the sizes of output of the two players that are random. For mathematical simplicity it is assumed that time t is continuous but the respective outputs are discrete integers. What we are interested in now is the joint probability distribution $F_{x,y}(t) = \text{Prob}\{X(t) = x, Y(t) = y; x,y = 0,1,2, \ldots, \infty\}$ of the output strategies of the two players and how it changes over time. Here we have a differential game with random moves by each player specified by the joint probability distribution function $F_{x,y}(t)$. As the reaction function in a deterministic Cournot model already maximizes the profit function of a given player, given a conjecture of the other's strategy, the distribution function $F_{x,y}(t)$ and its change over time incorporate implicitly a stochastic optimization by each player.

If we now assume that a bivariate Markovian (birth-and-death process) holds, then the following set of differential equations otherwise known as the Chapman-Kolmogorov equations [14] characterize the time-dependent evolution of the joint distribution of the two outputs:

$$\frac{dF_{x,y}(t)}{dt} = -(\lambda_x + \mu_x + \lambda_y + \mu_y)F_{x,y}(t) + \lambda_{x-1}F_{x-1,y}(t) + \mu_{x+1}F_{x+1,y}(t) + \lambda_{y-1}F_{x,y-1}(t) + \mu_{y+1}F_{x,y+1}(t) \qquad (7.1)$$

for x,y = 0,1,2, ..., ∞. The joint distribution function is defined to be zero for either x or y negative or both negative. Here λ_x (or λ_y) is the probability of a unit increase in the output of player one (or, two) given that it has attained a level x (or, y) at time t. Similarly μ_x, μ_y are the respective probabilities of a unit decrease in output for the two players. The λ's and μ's are also called birth and death rates, which are not fixed constants in this model but depend on the levels of output attained. This makes the model nonlinear in the variables and the parameters.

The Markovian specification (7.1) simply says that the rate of change of the joint distribution of the two outputs at t depends on three transition probabilities: a unit increase, a unit decrease or staying the same. For the Cournot model the parameters would reflect rivalry e.g.,

$$\lambda_x = \lambda_1 x, \quad \mu_x = \alpha_{11} x^2 + \alpha_{12} xy$$
$$\lambda_y = \lambda_2 y, \quad \mu_y = \alpha_{21} xy + \alpha_{22} y^2 \quad (7.2)$$

where λ_1, λ_2, α_{11}, α_{12}, α_{21} and α_{22} are fixed constants. Substitution of (7.2) in (7.1) leads to a set of two difference-differential equations for the dynamic behavior of the joint output distribution. These stochastic differential equations, which have their analogues in the deterministic framework can be utilized in two different ways so as to understand the dynamic behavior of a stochastic conjectural equilibrium. One is to apply the method of moments [15] so as to transform the stochastic differential equations above into equivalent equations in the expected values of X(t) and Y(t). Denoting these expected values by $m_1(t)$ and $m_2(t)$, their variances by $\sigma_1^2(t)$, $\sigma_2^2(t)$ and their correlation by $\rho(t)$ we may write the transformed system as:

CHAPTER III

$$\dot{m}_1(t) = \lambda_1 m_1(t) - \alpha_{11}(\sigma_1^2(t)+m_1^2(t))$$
$$-\alpha_{12}(\rho(t)\sigma_1(t)\sigma_2(t)+m_1(t)m_2(t)) \qquad (7.3)$$
$$\dot{m}_2(t) = \lambda_2 m_2(t) - \alpha_{21}(\rho(t)\sigma_1(t)\sigma_2(t)+m_1(t)m_2(t))$$
$$-\alpha_{22}(\sigma_2^2(t)+m_2^2(t))$$

Here it is assumed that the random moves are drawn from a fixed bivariate distribution for each t, so that the expected quantities $m_1(t)$, $m_2(t)$ satisfy the differential equations above, where dot denotes the time derivative. Such games with random moves have been considered by Restrepo [13], where the strategies of each player are functions of the outcome of a random move. These strategies are designed to fit a situation where, having received some information about the outcome of the random move, each player may select one out of a fixed distribution function such as $F_{x,y}(t)$, the number of alternatives being the same for both players. It is clear that the first player may not know $m_2(t)$ completely, although its conjecture $\hat{m}_2(t)$ may be made from past information. Likewise $\hat{m}_1(t)$ may be only known as conjectures by the second player. As before these conjectures variables may be used to define a dynamic conjectural equilibrium for given initial values $m_1(0)$, $m_2(0)$ as follows: A pair $\hat{m}_1(1)$, $\hat{m}_2(t)$ of conjectures is an equilibrium if it satisfies the consistency condition $\hat{m}_i(t) = m_i(t)$, i=1,2 and also solves the two nonlinear differential equations (7.3). Note that even if the correlation coefficient is zero, the differential system (7.3) above would involve interdependence e.g.,

$$\dot{m}_1 = \lambda_1 m_1 - \alpha_{11}(\sigma_1^2+m_1^2) - \alpha_{12} m_1 m_2$$
$$\qquad (7.4)$$
$$\dot{m}_2 = \lambda_2 m_2 - \alpha_{21} m_1 m_2 - \alpha_{22}(\sigma_2^2+m_2^2)$$

where α_{12}, α_{21} are not zero and the argument t has been suppressed. It is easy to see that these represent nonlinear reaction functions in a dynamic Cournot model, since at the steady state $\dot{m}_1 = 0 = \dot{m}_2$, these reduce to

$$m_1^2 = m_1 \left(\frac{\lambda_1}{\alpha_{11}} - \frac{\alpha_{12}}{\alpha_{11}} m_2 \right) - \sigma_1^2$$

$$m_2^2 = m_2 \left(\frac{\lambda_2}{\alpha_{11}} - \frac{\alpha_{21}}{\alpha_{11}} m_1 \right) - \sigma_2^2$$
(7.5)

where the variables are all time-independent steady-state values. For the fixed set of parameters $\theta = (\alpha_{11}, \alpha_{12}, \alpha_{21}, \alpha_{22})$, let $m_1^*(\theta)$, $m_2^*(\theta)$ solve this system (7.5) for any fixed σ_1^2, σ_2^2.

Theorem 2

If for any solution $m_i(t)$ of the differential system (7.4) the following two conditions hold for $i=1,2$

(a) if $m_i(t) > m_i^*(\theta)$, then $f_i(m_1, m_2) > \lambda_i$

(b) if $m_i(t) < m_i^*(\theta)$, then $f_i(m_1, m_2) < \lambda_i$

where $f_i(m_1, m_2) = \alpha_{ii}(m_i + \frac{\sigma_i^2}{m_i}) + \alpha_{ij} m_j$, $i \neq j$, $i,j=1,2$, then the steady state solution $m_j^*(\theta)$ has two-sided stability.

Proof

Under condition (a) we have $\dot{m}_i < 0$ around $m_i^*(\theta)$ and since any solution $m_i(t)$ of the differential equation system (7.4) is continuous in t, $m_i(t)$ would decline at t increases. Let $V(t) = \tfrac{1}{2}(m_i(t) - m_i^*(\theta))^2$ be the Lyapunov function. Then $\dot{V}(t) < 0$ for both conditions (a) and (b). Hence $m_i(t)$ converges to $m_i^*(\theta)$ under the above conditions. Hence the result.

Remark 2.1

If the nonlinear terms m_i^2 and $m_1 m_2$ are linearized around the steady state $m_i^*(\theta)$, when the latter is stable, then there exist a domain in the $(m_1(t), m_2(t))$-space where the Cournot equilibrium satisfies a pair of linear reaction functions.

Remark 2.2

There may be domains depending on the parameters θ and σ_i^2, where the steady state strategies $m_1^*(\theta)$, $m_2^*(\theta)$ may not be stable In such cases the players may not have any incentive to strive for such steady state strategies.

Remark 2.3

Expectations in this model may be interpreted subjectively by each player as in the dynamic entry-exit model analyzed by Brock [16]. If the actual size of output at time t is equal to the expected size, the forecast is then said to be an equilibrium forecast. The stochastic process model (7.1) and (7.2) may then be viewed as an adjustment process so as to reach an equilibrium forecast. This is very similar to the tâtonnement process applied to quantity adjustment.

A second way of interpreting the model specified by (7.1) and (7.2) is to compare the expected value analogue (7.4) with a deterministic formulation with two outputs now denoted as $x_1(t)$ and $x_2(t)$:

$$\dot{x}_1(t) = \lambda_1 x_1(t) - \alpha_{11} x_1^2(t) - \alpha_{12} x_1(t) x_2(t)$$
$$\dot{x}_2(t) = \lambda_2 x_2(t) - \alpha_{21} x_1(t) x_2(t) - \alpha_{22} x_{22}^2(t)$$
(7.6)

The steady state solution (x_1^*, x_2^*) now can be explicitly computed as:

$$x_1^* = (\lambda_1 \alpha_{22} - \lambda_2 \alpha_{12})/D, \quad x_2^* = (\alpha_{11}\lambda_2 - \alpha_{21}\lambda_1)/D$$

where $D = \alpha_{11}\alpha_{22} - \alpha_{12}\alpha_{21}$. The conditional stability of any $x_i(t)$ given that $x_j(t)$ is at its steady state level x_j^* ($j \neq i$) can also be easily solved as follows:

$$x_i(t) = (\frac{g_i}{\alpha_{ii}})[1 + (\frac{g_i}{\alpha_{ii} x_{i0}} - 1) \exp(-g_i t)]^{-1} \tag{7.7}$$

where $x_{i0} = x_i(t)$ at $t=0$, $g_i = \lambda_i - \alpha_{ij} x_j^*$ for $i,j=1,2$ and $i \neq j$. Thus it is clear $x_i(t)$ would grow if $\lambda_i < \alpha_{ij} x_j^*$ but would converge to x_j^* if $\lambda_i > \alpha_{ij} x_j^*$. Thus one could prove the following result:

Theorem 3

If the parameters λ_i, λ_j satisfy the ineuqality

$$\lambda_i/\lambda_j > (\alpha_{ii}\alpha_{ij})/(D + \alpha_{ij}\alpha_{ji}) \tag{7.8}$$

where $i \neq j$, $D = \alpha_{ii}\alpha_{jj} - \alpha_{ij}\alpha_{ji}$, then each output $x_i(t)$ converges for $t \to \infty$ to its steady state value x_i^* in a conditional sense.

Remark 3.1

In the general case when we do not restrict ourselves to conditional stability, the domain of stability may be more restricted, although it is true that for α_{12} and α_{21} zero, the model has steady state values that are stable.

CHAPTER III

Remark 3.2

In case the inequality in (7.8) is reversed, the convergence fails to hold but the outputs cannot grow to infinity either due to the finite total market demand or to suitable transversality conditions.

Two points are clear when we compare the deterministic system (7.6) with its expected value analogue (7.4). First, the former model shows a negative interdependence as is typical in Cournot reaction curves or lines, implying that the output size of one player varies inversely with that of his rival. This corresponds to a negative correlation (i.e., $\rho(t)<0$ in (7.3)) of the two outputs in the stochastic case (7.3). Thus the first interaction term $\alpha_{ij}\sigma_i(t)\sigma_j(t)$ has a positive impact on $\dot{m}_i$. Secondly, since $E\{x(t)y(t)\} < m_1(t)m_2(t)$ due to negative correlation, the solutions of the two models, the stochastic (7.4) and the deterministic one (7.6) cannot be identical unless α_{11} and α_{22} are zero. Hence we can prove the result:

Theorem 4

If the interaction coefficients α_{11}, α_{22} are zero in the model (7.6), then there exists a domain where the steady state solution (x_1^*, x_2^*) is necessarily stable and $x_i(t) = m_i(t)$ for $i=1,2$ and all t.

Proof

On using the Lyapunov function $V = \frac{1}{2}(x_1(t) + x_2(t) - (x_1^*+x_2^*))^2$ and taking its time derivative we get

$$\dot{V} = (x_1(t)+x_2(t)-\lambda_1-\lambda_2)(x_1(t)(\lambda_1-x_2(t)) - x_2(t)(\lambda_2-x_1(t)))$$

and this is negative for the two domains: (i) $x_1(t) > \lambda_2$, $x_2(t) > \lambda_1$ and (ii) $x_1(t) < \lambda_2$, $x_2(t) < \lambda_1$. Hence by Lyapunov's theorem there exists a domain where $x_1^* = \lambda_2$, $x_2^* = \lambda_1$ is stable.

Remark 4.1

Since $x_i(t) = m_i(t)$ for all t, the principle of certainty equivalence in the sense $E\{X(t)\} = x_1(t)$, $E\{Y(t)\} = x_2(t)$ holds. But if any of the two coefficients α_{11}, α_{12} are nonzero, the certainty equivalence does not hold.

Although we have so far discussed conjectural equilibrium in a duopoly context, it is not difficult to generalize it to the case of n players. We consider a very simple example where each player reacts by increasing (decreasing) his output when his expected profit $E(\pi_i(t))$ exceeds a minimal positive return $\bar{\Pi}_{i0}$ determined exogenously i.e.,

$$\dot{x}_i = k_i(E(\pi_i(t)) - \bar{\Pi}_{i0})$$
$$x_i(0) \text{ given}; \; i = 1,2,..,n \quad (8.1)$$

The market demand function is

$$p = a_0 - b \sum_{j=1}^{n} x_j + \varepsilon; \; E(\varepsilon) = 0, \; E(\varepsilon^2) = \sigma^2$$

hence the profit is

$$\pi_i(t) = (a - b \sum_{j=1}^{n} x_j) x_i$$

where $a = a_0 - c > 0$ and c_i is the marginal cost. For simplicity we consider a symmetric Cournot game where $c_i = c$, $k_i = k$ and $\bar{\pi}_{i0} = \bar{\pi}_0$ and there are no capacity constraints. The payoff function of each player is

CHAPTER III

$$\max_{x_i(t)} J_i = \int_0^\infty [\exp(-rt)\{E(\pi_i(t)) - \frac{w}{2} \text{var}(\pi_i(t))\}dt \qquad (8.2)$$

where the expected profit is $E(\pi_i(t)) = (a_0 - bX(t))x_i(t)$ with $X(t) = \sum_{j=1}^{n} x_j(t)$ and the variance of profit is $\text{var}(\pi_i(t)) = \sigma^2 x_i^2(t)$. Let $\hat{X}(t)$ be the conjecture by player i of other players' strategies in the form $\hat{X}(t) = x_i(t) + \sum_{j=1}^{n} \hat{x}_j(t)$, $j \neq i$ then for these conjectures to be consistent we require that the following equations hold for each $i = 1, 2, \ldots, n$:

$$\dot{z} = r\, z(t) - a + b(\hat{X}(t) + x_i(t)) + w\sigma^2 x_i(t)$$
$$+ k\, z(t)(a - bx_i(t) - b\hat{X}(t))$$
$$\dot{x}_i = k\{(a - b\hat{X}(t))x_i(t) - \bar{\Pi}_0\} \qquad (8.3)$$

where $z(t)$ is the adjoint variable in the Hamiltonian H:

$$H = \exp(-rt)\{ax_i(t) - bX(t)x_i(t)$$
$$+ kz(t)(ax_i(t) - bX(t))x_i(t) - \bar{\Pi}_0\}$$

Also we must have $\hat{X}(t) = X(t) = \sum_{i=1}^{n} x_i(t)$ for the conjectural equilibrium. Thus if the conjectural equilibrium exists, one may easily derive from (8.3) the following equations for Cournot-Nash consistent conjectural equilibrium:

$$\dot{z} = (r - k(1 + \frac{b}{n})X(t) + ak)z(t) + (b(1 + \frac{1}{n}) + \frac{w\sigma^2}{n})X(t) - a$$

$$\dot{X} = k(aX(t) - bX^2(t) - n\bar{\Pi}_0)$$

The steady state solutions z^*, X^* satisfy

$$z^* = (ak+r-AX^*)^{-1}(a-BX^*)$$
$$X^* = (2b)^{-1}(a \pm (a^2-4bn\bar{\Pi}_0)^{\frac{1}{2}})$$

where $A = k(1+\frac{b}{n})$, $B = b(1+\frac{1}{n}) + \frac{w\sigma^2}{n}$. Let the positive square root of $(a^2-4bn\bar{\Pi}_0)$ be denoted by $a-2h$ for some positive h. The higher the n the greater the value of h. The steady state value X^* then reduces to h/b, implying that higher n will lead to greater aggregate output. For large n, the steady state value z^* converges to:

$$z^* = \frac{a-bX^*}{r+ak-kX^*} = \frac{a-h}{r+k(a-\frac{h}{b})}$$

Note that this z^* is independent of the risk aversion term ($w\sigma^2$).

3. Stabilization Policy in Cartelized Markets

Recent theoretical discussions of commodity stabilization policy through buffer fund and/or international commodity agreements have stressed two new aspects of the policy problem which are of great importance for the less developed countries (LDCs) and the North-South trade. The first aspect refers to the role of dominant producers in imperfect or cartelized markets, which may be significantly different from the competitive frameworks. Thus Newbery [17] has developed the hypothesis that if demand is linear, then dominant producers find it optimal to undertake significantly more storage than is competitively justified and this storage policy increases price stability in proportion to the market share of the dominant producers. The second aspect deals with the risk-sensitivity of buffer stock intervention rules and the extent to which conventional decision rules could be improved so as to protect the producers of primary commodities in LDCs against the uncertainty of future export earnings. Several features of this problem have been analyzed by Goreux [18], Nguyen [19] and Hughes-Hallett [20]. The contributions by Nguyen and Hughes-Hallett specifically introduce the need for minimizing the variance of export earnings as a goal to be pursued by the buffer fund authority and in the framework of linear quadratic Gaussian (LQG) model of optimal control this introduces robustness into a buffer stock policy. This has been stressed by Whittle [21] as a risk-sensitive generalization of the traditional LQG optimal control rule. This may be specially important for some monopolistic commodity markets like coffee and cocoa, where the suppliers may be more sensitive to the risk of fluctuations in price and profits.

Our objective here is two-fold: to examine the relevance of the non-cooperative game model in world commodities like coffee and its implications for an integrated stabilization policy on the lines of the Integrated Programme objectives of UNCTAD formulated in 1976. Two main theoretical issues are posed here. One is the difficulty of implementing a robust stabilization policy in an oligopolistic stochastic market, where the role of supply uncertainty and market information is usually asymmetrical. The second is to evaluate the impact of instability associated with such non-cooperative game-theoretic models, whereby the transmission of such instabilities to the LDCs may affect their development and diversification goals. This aspect of transmission of instabilities, when viewed in the light of the major primary commodities in world trade may be very substantial.

The world coffee market, with its significant price and quantity fluctuations similar to that of copper provides in many ways a unique framework for testing the empirical realism of the hypothesis of dominant producers playing a significant role. The main reason is that Brazil has been and still is, the world's largest single coffee-producing country although its share of the world market has been steadily declining from over 60% in the 1940's to around 30% in the late 1970's. The share of world inventories held by Brazil declined from 90% or more around 1962 and before, to about 35% or less around 1980. A second unique feature of the world coffee economy is that Brazil has contributed very significantly to the year to year fluctuations in world production over the years 1946-1967 for which reliable estimates are available.

These considerations lead very naturally to the specification of a non-cooperative game-theoretic model for the world coffee economy, which may be empirically tested. For this purpose two sample periods 1945-46 to 1961-62 and 1962-63 to 1979-80 have been considered for each coffee agricultural year (July to June). Sample I starts with the 1945 season in order to avoid the Second World War and 1962 was chosen to be the last year of Sample I, since an important ICA was signed then with a stronger enforcement of the export quota system. From a set of different forms of demand functions, the following turned out to be the best, in terms of R^2 and the plausible signs of the regression coefficients.

$$\text{Sample I} \quad p(t) = 29.42 - 0.0026^* x_1(t) + 1.988(10^{-8})^* x_1^2(t)$$
$$(1945\text{-}1962) \quad \quad (0.90) \quad (-2.49) \quad \quad (2.01)$$

$$- 0.0002\, x_2(t) + 0.034^*\, y(t) \quad \quad (9.1)$$
$$(-0.25) \quad \quad (2.16)$$

$$R^2 = 0.87; \quad DW = 2.15$$

CHAPTER III

Sample II
(1962-80)

$$p(t) = 455.32^{**} - 0.0067^* \, x_1(t) + 4.01(10^{-8}) \, x_1^2(t)$$
$$(3.2) \quad\quad (-2.15) \quad\quad\quad (1.61)$$

$$- 0.0040^* \, x_2(t) + 0.012 \, y(t) \quad\quad\quad (9.2)$$
$$(-2.07) \quad\quad\quad (1.34)$$

$$R^2 = 0.87; \; DW = 1.88$$

Here $p(t)$ is the New York price of Brazilian coffee, $x_1(t)$, $x_2(t)$ are the supply quantities in year t of Brazil and the rest of the world (RW), where supply is defined as current production plus the inventory held at the beginning of the current year; $y(t)$ is the U.S. gross national product per capita in real terms. The t-statistics are given in parentheses, one asterisk denoting significance at 5% level and two at 1% level. Some comments are in order about the two empirical estimates presented in (9.1) and (9.2). First, we note that these demand functons lead to Cournot-Nash reaction functions, if we add the usual profit-maximization assumption and zero conjectural variation. It is true there are more than two major producers in the world coffee economy, but Brazil being the largest producer can be assumed as one duopolist, while the rest of the world (RW) may be treated as a rival. The behavior of all coffee producing countries other than Brazil may not exactly be similar to that of a single producer, but it seems reasonable to presume that Brazil may consider other producers as a potential threat to its dominant market share and thus consider them as a rival group. Moreover due to multicollinearity, many rivals cannot easily be introduced in the supply equation estimation. Second, the role of lagged inventory is very important in each of the two demand equations. For example if the supply quantity is replaced by production ($q_i(t)$) only, we obtain the following linear estimates (nonlinear terms were insignificant):

Sample I: $p(t) = -9.45 - 0.0011 \, q_1(t) - 0.0029^* \, q_2(t)$
             $(-0.76) \, (-1.98) (-2.12)$

$ + 0.0691^{**} \, y(t)$ \hfill (9.3)
$ (4.45)$

$ R^2 = 0.73; \, DW = 2.01$

Sample II: $p.(t) = 244.46 - 0.0015 \, q_1(t) - 0.0071^* \, q_2(t)$
             $(1.88) (-1.31) (-2.08)$

$ + 0.0372^{**} \, y(t)$ \hfill (9.4)
$ (5.09)$

$ R^2 = 0.82; \, DW = 2.1$

It is clear that Brazil's production level has no influence on price that is statistically significant at the 5% level, although it is a dominant producer in the world market.

To derive Cournot-Nash (CN) equilibrium solutions from the estimated demand functions (9.1) and (9.2), one needs annual average cost data for the two sample periods. Since this is unavailable, we take an average estimate of 2.5% as storage costs and 77.5% as production costs as percentages of wholesale price in 1945 and for subsequent years cost was adjusted at the same rate as wholesale prices. This method is consistent with the estimates made by Geer [22], who noted that the additional production capacities created in RW in the 1960's reflected in part low production costs relative to the maintained price in world market. The optimal reaction curves satisfying the first order conditions of a CN equilibrium turn out as follows with k_i denoting the demand effect of $y(t)$:

Sample I: $x_1^* = 5{,}503 + k_1 - 0.0385 \, x_2^* + 0.00001 \, x_1^{*2}$
             $x_2^* = 71{,}550 + k_2 - 6.5001 \, x_1^* + 0.00005 \, x_1^{*2}$ \hfill (9.5)

and,

Sample II:
$$x_1^* = 33,880 + k_1 - 0.2985\ x_2^* + 0.00001\ x_2^{*2}$$
$$x_2^* = 56,000 + k_2 - 0.8401\ x_1^* + 0.000005\ x_1^{*2}$$
(9.6)

The second order conditions for maximum profits require that $x_1^* < 43,000$ (Sample I) and $x_1^* < 53,600$ (Sample II). Two results are most significant here. One is that in sample period II the influence of RW on Brazil's optimal supply has increased - by almost eight times, i.e., $\partial x_1^*/\partial x_2^*\ _{II} = 7.75\ \partial x_1^*/\partial x_2^*\ _I$. The corresponding impact of Brazil on RW has declined considerably, e.g., by 0.13 times. Second, the nonlinear reaction functions show very clearly that in the space of optimal supply, x_1^* and x_2^* are not always negatively correlated. This implies that correlation of strategies may at times be deliberately utilized by Brazil as a tool to improve the non-cooperative CN equilibrium solution in such market games. In general however when the correlation of x_1^* and x_2^* is negative, the nonlinear elements in (9.5) and (9.6) augment the speed of convergence to a stable CN equilibrium solution. The final equilibrium solutions predicted by the optimal reaction curves (9.5) and (9.6) do however overestimate the average supply behavior observed during the two sample periods. Two reasons for this discrepancy may be given. One is that the CN models above fail to incorporate future expectations about trends in supply and demand. The second reason is that the reaction curves fail to incorporate both supply (yield) and demand (price) risks. This contributes in part to the overestimate of optimal supply in relation to the observed actual supply. Some idea of the commodity price cycles may be obtained from the estimates (1953-74) made by Reynolds [23] as follows:

	complete cycle (peak to peak)		downswing only (peak to trough)	
	average length (months)	max. length	average length	max. length
coffee	34	79	18	45
cocoa	17	33	8	15
sugar	13	26	7	20

Although no direct estimate of the yield risk are available along similar lines, the fluctuations in coffee production during the same period have been significantly high. According to one estimate [19], the fluctuation index computed as an average over 1953-74 of differences between annual observations and calculated trend values expressed as percent of trend value was as follows:

Exports from LDC
Indices of fluctuations

	market price	value
coffee	22.1	15.7
cocoa	33.9	18.6
sugar	54.2	20.0

In addition to production fluctuations, world producer stocks of coffee show a long term cycle that fairly accurately mirrors the price cycle, with an all-time low in the mid-1950's, very high levels during the 1960's and lower in the early 1970's. The main holder of coffee stocks among producers till 1970 has always been Brazil, due to the fact that it is the only country which could actually increase its export earnings by reducing its quantity of supply in the world market. To explain these fluctuations we need a modified CN model as follows.

CHAPTER III

The modified CN model allows for a learning mechanism and the elements of risk aversion, when there are n (n $\geq$ 2) suppliers in a market. For the i-th supplier the profit $\Pi_i(t)$ at time t can be written as

$$\Pi_i(t) = R_i(t) - C_i(t) \qquad (9.5)$$

where
$$R_i(t) = (\hat{p} + u) x_i(t) \quad \text{(revenue)}$$
$$C_i(t) = (c + v) x_i(t) \quad \text{(cost)}$$
$$\hat{p} = E(p(t+1)|S(t-1)) \qquad (9.6)$$
$$= a_0 - b\, S(t-1) + \beta\, S^2(t-1) \quad \text{(expected price)}$$

$$S(t) = \sum_{j=1}^{n} x_j(t) \quad \text{(aggregate supply)}$$

Here the simplifying assumptions are that both expected price and expected marginal cost are similar across suppliers, u and v are random disturbance terms normally and independently distributed with zero means and finite variances σ_u^2, σ_v^2 which denote price and yield risks respectively. Each supplier is assumed to forecast $\hat{p}$, i.e., the level of expected price one period ahead and then choose his own optimal supply $x_i^*(t)$ by maximizing the expected utility $f_i = E\, U(\Pi_i(t))$ of profit where the utility function is of the exponential form

$$U = 1 - \exp(-\alpha_i \Pi_i(t)), \quad \alpha_i > 0 \qquad (9.7)$$

with a constant rate of absolute risk aversion denoted by α_i. This leads to the following maximand for each CN player

$$\max f_i = [\, a - b\, S(t-1) + \beta\, S^2(t-1)\,]\, x_i(t) - \tfrac{1}{2} \alpha \sigma^2 x_i^2(t) \qquad (9.8)$$
$$\text{where } a = a_0 - c,\ \sigma^2 = \sigma_u^2 + \sigma_v^2$$

Note that this maximand, though myopic in perspective seems to be consistent with the two-year coffee-bearing cycle noted by Geer and others [22,23]. It is clear from (9.8) that the CN equilibrium solutions would satisfy the following nonlinear difference equation

$$S(t) = \theta_0 - \theta_1 S(t-1) + \theta_2 S^2(t-1) \qquad (9.9)$$
$$\text{where } \theta_0 = (na)/(\alpha\sigma^2), \ \theta_1 = (nb)/\alpha\sigma^2)$$
$$\theta_2 = (n\beta)/(\alpha\sigma^2)$$

One has to note two important implications of the difference equation. First, consider a linear version by ignoring the nonlinear term: the solution then appears as

$$S(t) = [S(0) - (\theta_1)^{-1} \theta_0](-\theta_1)^t + (\theta_0/\theta_1)$$

The cycles represented by the range of $[(-\theta_1)^t, t = \text{odd}, (-\theta_1)^t, t = \text{even}]$ clearly increases for increased $t > 1$, if either (nb) increases or, $(\alpha\sigma^2)$ decreases. Since the increase of competition consequent on the decline in dominance of Brazil as the largest producer is most likely to lead to reduce $\alpha\sigma^2$ or, increase (nb), it is clear that the range of fluctuations is likely to increase. Since the level of producer stocks held under competitive conditions would be much less than under imperfect competition under a dominant producer, the source of fluctuations may increase due to the existence of the so-called 'thin market' that has been stressed by Geer [21]. Secondly, the impact of the nonlinear term $\theta_2 S^2(t-1)$ is to decrease the range of fluctuations, since by linearly approximating it as $2\theta_2 \bar{S} S(t-1)$, where $\bar{S}$ is the mean value it changes the transient term $(-\theta_1)^t$ to $(-\theta_1 + 2\theta_2 \bar{S})^t$. However the nonlinear element may contribute to oscillations due to complex roots of the associated quasi-polynomial (i.e., quasi-characteristic equation). To see its implication we may use the linearizing approximations $\dot{S}(t) \sim S(t) - S(t-1)$ and $S^2(t-1) \sim \bar{S}^2 + [S(t-1) - \bar{S}](2\bar{S})$, whereby the homogeneous part of (10.2) reduces to the following system

$$\dot{S}(t) = \theta_3 S(t-1) + \theta_4 \qquad (10.1)$$
$$\theta_3 = 2\theta_2 \bar{S} - \theta_1 - 1$$
$$\theta_4 = \theta_0 - 2\bar{S}^2$$

On using another transformation $S(t) = z(t) \exp(rt)$ this finally reduces to the linear difference-differential equation (homogeneous part only):

$$\dot{z}(t) + r\, z(t) + \theta_5\, z(t-1) = 0 \qquad (10.2)$$
$$\theta_5 = -\theta_3 \exp(-r)$$

The characteristic quasi-polynomial for (10.2) is

$$\phi(\lambda) = \lambda + r + \theta_5 \exp(-\tau\lambda), \quad \tau = 1 \qquad (10.3)$$

with $\phi(\lambda) = 0$ being the characteristic equation. Various regions of the solution space characterized by the real and the complex values of λ may be characterized by the so-called D-partition method. Thus the equation $\phi(\lambda) = 0$ clearly has a zero root for $r + \theta_5 = 0$, which is a straight line forming the boundary of one region of the D-partition. Now let the quasi-polynomial have the purely imaginary root iq:

$$iq + r + \theta_5 \exp(-\tau\, iq) = 0; \quad i = +\sqrt{-1}$$

or

$$iq + r + \theta_5 (\cos \tau q - i \sin \tau q) = 0$$

On separating the real and imaginary parts and solving for r and θ_5 we get

$$\theta_5 = q(\sin \tau q), \quad r = (-q \cos \tau q)(\sin \tau q)^{-1} \qquad (10.4)$$

where

$$C: \{r + \theta_5 \cos \tau q = 0, \; q - \theta_5 \sin \tau q = 0\} \qquad (10.5)$$

defines a curve on which all the roots λ of the quasi-polynomial (10.3) are purely imaginary. Clearly if $r > 0$ and $\theta_5 = 0$, then the degenerate quasi-polynomial (10.3) has no roots with positive real parts and hence the system (10.2) is then asymptotically stable. However there arises at least one root with a positive real part when $\theta_5 \tau$ exceeds zero. Furthermore, if $\theta_5 \tau$ exceeds unity, we get $\cos \tau q < 0$ on the curve C defined by (10.5). As a result we have unstable oscillations. It is clear that the Cournot-Nash equilibrium solutions in such cases would not possess any stability property that may be desirable by the players.

The above results on regions of asymptotic stability without oscillations and asymptotic instability with oscillations have two broad implications for stabilization policies. One is that the nonlinear elements in the CN reaction curves contribute to oscillations due to complex roots, whenever the supply lag τ tends to increase above zero. The higher the lag, the greater the asymptotic instability along the curve C defined in (10.5). The role of producer stock level in smoothing out this incidence of production lags of 5 years or more for coffee becomes very critical. This also makes the case for a cooperative solution with an appropriate level of world buffer stock much stronger.

Secondly, the existence of unstable regions in the space of eigenvalues explains in part the behavior of Brazil, when it did not fulfil its export quota obligation under ICA (International Coffee Agreement), although coffee prices started rising in 1964 and reached very high levels in 1969-70. During this period the share of African countries in world exportable production increased from 25% in 1960-61 to 25% in 1965-66 and to 38% in 1976-77. It is clear that the ICA of 1969 and 1976 sought, rather unsuccessfully, to bring about long term equilibrium between production and consumption through a system of variable export quotas. But the enforcement of quota control has posed a difficult problem due to two main reasons. One is that some producing countries invariably attempted to unload their surplus production to non quota markets thus leading to a two-price system and defeating the overall stabilization objective. Second, due to the lack of effective implementation of long term production control and diversification and very low price elasticity of demand (range: -0.105 to -0.233 over 1960-1975), the producing countries participating in the ICAs could not rationally believe in the cooperative goal by which their total export earnings could increase.

It is clear therefore that for a world buffer stock program to succeed, a long term program must be developed along with the short term stabilization objectives. Since the export earnings from commodities like coffee, cocoa, etc. comprise a very significant share of total export revenues of the LDC and since this has sizable impact on per capita national income and its growth in these countries, it may be more economic and rational to support an integrated program of common fund based on ten or so core commodities and then emphasize the quantum of gains to be obtained from a cooperative framework of stabilization.

It is important to realize that short-run stabilization policies have long run costs and benefits and thus the overall costs. As it has been stressed by Calmfors [24] in his introduction to an international symposium on stabilization policy, there are four main effects of stabilizatin policy: (a) effects on future consumption possibilities, (b) effects on real capital investment and hence on future growth, (c) effects on long run production structure and (d) the effects on future domestic price and wage-setting behavior. To these may be added the higher risk-sensitivity of producers in cartelized markets, where inventories play a more active role. To see the optimal control implications consider the objective function of the suppliers as a discounted profit function with a constant discount factor $\beta = 1/(1+r)$, where r is the exogenous rate of interest and profit is

$$\pi_t = p_t x_t - c(z_t) - c(h_t) \tag{11.1}$$

where random demand x_t has mean $\bar{x}_t$ and a constant variance and the inventory h_t is defined by

$$h_t = h_{t-1} + z_t - x_t; \quad h_t \geq 0 \tag{11.2}$$

The cost functions $c(z_t)$, $c(h_t)$ for production and inventories are assumed convex and differentiable, where primes denote the respective marginal costs. Further the monopolistic firm is now assumed to be risk averse, so that his objective funcion takes the form

$$\text{Maximize } E\{\sum_{t=1}^{T} \beta^t \{\pi_t - \frac{\lambda}{2}(R_t - ER_t)^2\} \tag{11.3}$$

where $R_t = p_t x_t$ is total revenue, E is expectation as of t and λ is a non-negative constant. This formulation is exactly similar to that of Newbery

CHAPTER III

[17] except for the risk aversion term. Newbery showed that the monopolistic and cartelized firms do carry much larger inventories than the competitive firms in order to reduce instability of profits. It is clear from the conditions above that along the optimal trajectory we must satisfy the following set of complementary inequalities:

$$c'(z_t) + \lambda \sigma_R (\partial \sigma_R / z_t) > m_t$$
$$m_t + c'(h_t) + \lambda \sigma_R (\partial \sigma_R / \partial h_t) \geq \beta E m_{t+1} \quad (11/4)$$
$$h_t \geq 0$$

By Kuhn-Tucker theorem these necessary conditions are also sufficient. Here the marginal costs are indicated by a prime, m_t is marginal revenue at time t and σ_R^2 is the variance to total revenue $R = p_t x_t$, when demand alone is assumed to be random. Assuming interior solutions, the above necessary conditions reduce to

$$\beta E m_{t+1} = c'(z_t) + c'(h_t) + c(h_t) + \lambda \sigma_R \left(\frac{\partial \delta_R}{\partial z_t} + \frac{\partial \sigma_R}{\partial h_t} \right) \quad (11.5)$$

For competitive firms the analogous condition is

$$\beta E m_{t+1} = c'(z_t) + c'(h_t) + \lambda \sigma_R \left(\frac{\partial \delta_R}{\partial z_t} + \frac{\partial \sigma_R}{\partial h_t} \right) \quad (11.6)$$

Two points are now clear. The risk-adjustment costs due to the variance of total revenue enter as the last term in both the optimality conditions (11.5) and (11.6) and in the risk-less case ($\lambda=0$). Secondly, the monopolisic firm would normally carry larger inventories than a competitive firm but for some t the price may decline for a monopolistic firm. Note that any forecsting error in predicting demand is reflected in the marginal terms $\partial \sigma_R / \partial z_t$ and $\partial \sigma_R / \partial h_t$ which is nothing but risk-sensitivity. In case of monopolistic or cartelized markets these risk-sensitive impacts may be much more significant than in competitive markets.

4. <u>Adaptive Controls in Economic Models</u>

Methods of adaptive control which are increasingly applied in engineering systems theory and economic fields [25,26], basically deal with some of the major problems in applying optimal control theory to real life situations. Some of these problems are as follows: (1) how would parameter uncertainty or, parameter variations affect the optimal control rules? (2) if the parameters are estimated, how would the estimation and control objectives be combined if at all? (3) how would one evaluate the various sub-optimal control rules obtained by suitably approximating the nonlinear optimal control rules, when the model is more nonlinear than an LQG framework? (4) and, how could one rely on an optimal control rule in a stochastic control model, when it implies either a bang-bang rule (i.e. alternating between two modes, one maximal and the other minimal) or, a fluctuating rule?

In engineering systems e.g., chemical process control two most widely adopted approaches to adaptive control are self-tuning regulators (STR) and model reference control. The former attempts to build adaptivity through combining the two problems of optimal estimation and optimal regulations, whereas the latter provides a comparison with a reference model such as the LQG model where an optimal feedback control rule exists, and thereby monitor the deviations from the reference model if they are very large or widely fluctuating. In economic fields two most widely used applications are the use of cautious control rules in uncertain parameter systems and, the stability aspects of suboptimal control rules in stochastic control models. The former arises when some of the parameters have to be statistically estimated over given and also evolving data on the state variables and hence the role of alternative information structures in updating the optimal or pseudo-optimal control rules is very important here [26]. The latter emphasizes the point that an optimal control model, which in its adjoint equations is deterministically stable may not be so, when the stochastic elements are introduced so as to transform the adjoint equations to a set of stochastic difference or differential equations.

We consider here three examples of adaptive control: (1) a cautious control, (2) a self-tuning regulator (STR) and (3) an optimal control in a randomly fluctuating environment.

4.1 Cautious Control Policy

The most widely used example of cautious control combines the payoffs from two apparently conflicting goals: good estimation and good regulation. Good estimation often requires a large spread in the instrument or control variables, since it would make the model more reliable and representative, whereas good regulatory control requires small variations in the instruments, since this would make the optimal policy fine tuned. As a scalar example consider one state variable (y_t) and one control (x_t) related by

$$y_t = \alpha + \beta x_t + \varepsilon_t; \quad t = 1, 2, \ldots \ldots \tag{12.1}$$

where ε_t is a zero-mean random variable with constant variance σ_ε^2 and it is independent of control. Since this satisfies the requirement of a least squares (LS) model, we assume that the LS estimates $(\hat{\alpha}, \hat{\beta})$ are available from past observations of the pair $(x_t, y_t; t=1,2,\ldots,K)$. This is the estimation part. Now for the part of optimal (or good) regulation or control, we have to choose a performance measure or a criterion function. On using a one-period loss function L for example

$$L = E(y_{t+1} - y^F)^2 \tag{12.2}$$

where E is expectation and y^F is a desired goal or target, and the dynamic model (12.1), one could easily compute the optimal control rule x^*_{t+1} say:

$$x^*_{t+1} = \hat{\beta}_t (y^F - \hat{\alpha}_t) \{\hat{\beta}_t^2 + \sigma^2(\hat{\beta}_t)\}^{-1} \tag{12.3}$$

where the LS estimates $\hat{\alpha}_t, \hat{\beta}_t$ are given the time subscript t, to indicate that it is computed over all observed data (x_s, y_s) available up to t i.e., $s=1,2,\ldots,t$. Here $\sigma^2(\hat{\beta}_t)$ denotes the squared standard error of the LS estimate $\hat{\beta}_t$ i.e.,

$$\sigma^2(\hat{\beta}_t) = \sigma^2_\varepsilon / \sum_{s=1}^{t} x_s^2 \tag{12.4}$$

Note the two major elements of caution in the optimal control rule (12.3). It has the squared standard error of $\hat{\beta}_t$ in the denominator, implying that a large standard error would generate a very low control x^*_{t+1}. This property would be much more magnified if we replace the one-period loss function by an intertemporal one e.g.,

$$L = E[\sum_{t=1}^{T}(y_{t+1}-y^F)^2] \qquad (12.5)$$

This is because we have to know or guess the future values of control as of t and hence that of k-step forecast of $\hat{\beta}_{t+k}$ and its standard error. The second element of caution is due to the gap of $\hat{\alpha}_t$ from the target level y^F. If the standard error of $\hat{\alpha}_t$ which is $\sigma_\varepsilon/\sqrt{t}$ is very large, then the estimate of this gap is also very unreliable, thus implying the less reliability of of the control rule x^*_{t+1}; also if the desired target rate y^F is set too high, it may require a large control, hence a large cost of control.

Moving the sample size from t (i.e., s=1,2,..,t) to t+k say one could iteratively compute $(\hat{\alpha}_t,\hat{\beta}_t)$ and then x^*_{t+1} up to t+k and then reestimate $\hat{\alpha}_{t+k}$, $\hat{\beta}_{t+k}$ by LS methods and so on. Some Monte Carlo simulation experiments reported by Taylor [27] and Stenlund [28] suggest that the successive estimates $(\hat{\alpha}_t,\hat{\beta})$ may be biased and sometimes heavily and the standard error of $\hat{\beta}_t$ may decrease to zero very slowly even under large sample conditions.

These two problems are likely to be magnified [26] in case the objective function is of intertemporal form (12.5). Also for multivariable cases with state and control vectors, the two-stage process of control and estimation may involve substantial computational load.

4.2 Self-tuning Regulator

The principle of self-tuning control has enjoyed a growing interest in both theory and applications since its appearance in the control literature in the early 1970s. The combination of a LS parameter estimator with a minimum variance regulator led to a self-tuning regulator (STR) of the direct type involving direct estimation of the control parameters. However as we indicated earlier there remained many computational problems, particularly the high degree of sensitivity of the minimum variance regulation. Some flexibility was introduced by Clarke [29] and several types of other generalizations have been recently discussed by Keyser and van Cauwenberghe [30]. Thus the direct type of STR originally discovered by Ljung and Astrom have led to other more indirect types of STR. We present here a simple example from Clarke [29] and then emphasize some of its generalized implications. In the economic field STR is important as an adaptive process because it shows that the estimation is much more a part of the optimal control problem.

Consider a scalar state or output variable y_t and a control variable u_t related by the dynamic system

$$y_t + a_1 y_{t-1} + \ldots + a_n y_{t-n}$$
$$= b_0 u_{t-k} + b_1 u_{t-k-1} + \ldots + b_n u_{t-k-n} \qquad (13.1)$$
$$+ \varepsilon_t + c_1 \varepsilon_{t-1} + \ldots + c_n \varepsilon_{t-n}$$

which may be rewirtten as

$$A(q^{-1})y_t = B(q^{-1})u_{t-k} + C(q^{-1})\varepsilon_t \qquad (13.2)$$

where A,B,C are polynomials in q^{-1}, and u_t, y_t are deivations from the nominal mean levels $\bar{u}$, $\bar{y}$ and ε_t is an uncorrelated random sequence with zero mean and constant variance σ^2. Following the method of Astrom and Wittenmark [31] we choose the control objective as one of minimizing the variance of output y_t. To derive the STR algorithm we write (13.2) in a predictor form

$$y_{t+k} = \frac{B}{A} u_t + q^k \frac{C}{A} \varepsilon_t \qquad (13.3)$$

where the argument q^{-1} is omitted in polynomials A, B and C. On using the polynomial identity

$$\frac{C}{A} = G + q^{-k}\frac{F}{A} \qquad (13.4)$$

where $G(q^{-1})$ is a polynomial of order k-1 we obtain after some simplification

$$y_{t+k} = \frac{Fy_t + GBu_t}{C} + G\varepsilon_{t+k} \qquad (13.5)$$

$$= \hat{y}_{t+k|t} + \tilde{y}_{t+k|t} \qquad (13.6)$$

where $\hat{y}_{t+k|t}$ is the k-step ahead prediction of output as of time t and $\tilde{y}$ is the prediction error which is independent of $\hat{y}$, since all terms in ε contain future errors in the form ε_{t+j}, $1 \leq j \leq k$. Hence the variance of output, $E(y_{t+k}^2)$ is minimized by choosing u_t to set $\hat{y}_{t+k|t}$ equal to zero. In this case the minimum output variance becomes

$$E(\tilde{y}^2) = \sigma^2(1 + e_1^2 + \ldots + e_{k-1}^2)$$

Thus the STR algorithm works as follows: The explicit form of the self-tuning control could use a generalized least squares method to estimate the polynomials A, B and C and then use the identity of (13.4) to derive the parameters G and F that are used in the optimal control rule equation (13.5). Under certain regularity conditions e.g., a stable open-loop control system, Astrom and Wittenmark proved that such a two-stage process converges and the STR then defines an optimal controller with minimum variance of output.

Note that instead of the explicit form of STR, one could use an implicit form also, where we first use a regression model of the form

$$y_t = Fy_{t-k} + Hu_{t-k} + e_t; \quad H = GB$$

to estimate the parameters of the F and H polnomials.

The objective of the STR proposed by Astrom and Wittenmark [31[i.e., to minimize the output variance is however very restrictive, since it cannot prevent large fluctuations of sudden changes in output levels. Clarke and Gawthrop [32] proposed a more generalized loss function

$$\text{Min } L, \quad L = E\{\phi_{t+k}^2 | t\}$$

where

$$\phi_{t+k} = Py_{t+k} + Qu_t - w_t \qquad (13.7)$$

ϕ_{t+k} is a generalized form of output defined in (13.7), where $P = P(q^{-1})$, $Q = Q(q^{-1})$ are suitable transfer functions that can be specified by the control designer to achieve a variety of goals and the signal w_t may include tracking and regulatory goals.

Recent generalizations of the STR in both direct (i.e., explicit) and indirect (i.e., implicit) forms have developed algorithms having a very important property: the parameter estimations are not directly dependent on the implemented control structures. This is because the estimation part of the STR is essentially related to the prediction problem and not to the control problem.

The STR processes and the associated algorithms hold great promise for economic models for two main reasons. One is that it can provide valuable insights into the learning process of one or many policymakers. This may be important for economic models of rational expectations [26]. Even when the STR process does not converge, it may suggest the reasons why the equilibrium assumptions of the model specification may not all hold. A second reason is that the STR process when it exists may imply a certain type of robustness property that is very helpful in the design and implementation of macroeconomic policy. For economic applications one may explore generalizing the least squares method of estimation to include other types of nonnormal distribution of errors; also the problem of smallness of the sample size in economic time series has to be tackled through suitable generation of artifical simulated data.

4.3 Optimal Control in Random Environments

The problem of specifying or designing an optimal control in a fluctuating random environment may be viewed in several ways of which the following two would be illustrated: one provides for smoothing out the fluctuations, while the second incorporates a learning mechanism with a probabilistic reinforcement scheme.

The first type of situation arises in ecology where the problem is how to optimally harvest a randomly fluctuating population. A number of recent models has been surveyed by Mangel [33] in this connection. We consider here one such model studied by Gleit [34] in the continuous time case and by Mendelssohn [35] in the discrete case. Consider a fishing population characterized by a scalar stochastic variable $X = X(t)$, whose dynamics of growth and fluctuations is specified by the following stochastic differential equation

$$dX = [f(X,t) - h(X,t)]dt + \sigma(X,t)dW \qquad (14.1)$$
$$\text{for } X > 0$$

Here $f(X,t)$ is the natural deterministic growth rate, $h(X,t)$ the rate of harvest, $\sigma(X,t)$ a measure of the intensity of noise and $W(t)$ a standard Brownian motion stochastic process with expectaton $E\{dW\} = 0$ and $E\{(dW)^2\} = dt$. The profit from a harvest rate $h(X,t)$ is taken as

$$\pi(X,t,h(X,t)) = [p(t) - c(X,h,t)]h(X,t)$$

where $p(t)$ = price per unit harvest, $c(X,h,t)$ is the cost of harvesting a unit biomass when the stock size is X and the harvest rate is h. A long run expected value of discounted profits may then be set up as the objective function to be maximized:

$$\max_{h(X,t)} J = E\{\int_0^T e^{-rt}\pi(X,t)dt + e^{-rT}S(X(T),T)\}$$

where $S(X(T),T)$ is the preservation value of stock at time T and the exogenous rate of discount is $r>0$. By using the standard dynamic programming algorithm by setting

$$J(x,t) = \max_h E\{\int_t^T e^{-rs}\pi(X,s)ds + e^{-rs}S(X(T),T\ X(t) = x\}$$

we obtain the recursive equation by the principle of optimality:

$$0 = J_t + \max_h\{\tfrac{1}{2}\sigma^2(x,t)J_{xx} + [f(x,t)-h(x,t)J_x + e^{-rt}\pi\} \qquad (14.3)$$

where the subscripts of J denote the partial derivatives. This nonlinear differential equation (14.3) has to be solved for computing the optimal sequence of harvest. In some special cases the solution can be explicitly written in a closed form e.g., let the cost function c(x) depend on x only and the harvest rate be independent of x, then one obtains from (14.3) the optimality condition

$$J_x = e^{-rt}[p(t) - c(x)]$$

from which the optimal value of h can be computed. However even in this case the optimal control rule h may take the form of a bang-bang control i.e., of the form

$$h = \begin{cases} h_{max} \\ h_{min} \end{cases}$$

which is socially very undesirable, since these controls are essentially "boom or bust" and due to extensive capitalization requirements they are very expensive. One way to smooth out these bang-bang type fluctuations is to modify the objective functional so that we have the following model:

$$J(x,t) = \max_h E\{\int_t^\infty [(p-c(x))h - \beta(\frac{dh}{dt})^2]e^{-rs} \, ds \,|\, X(t) = x\}$$

and

$$dX = \{f(x) - h\}dt + \sigma(x)dW$$

where it is assumed $f(X,t) = f(X)$, $\sigma(X,t) = \sigma(X)$ and β is a parameter measuring the cost of changing effort.

Mendelssohn [35] adopts a similar procedure to smooth out the widely fluctuating harvest rates resulting from the bang-bang tupe optimal controls but in a discrete-time system. Let x_t be the population size in year t before the harvest z_t and $y_t = x_t - z_t$. The dynamics of the stock is given by

$$x_{t+1} = f(y_t, \varepsilon_t) \qquad (15.1)$$

where ε_t are independent, identically distributed random variables and $f(\cdot)$ is the growth function. The objective function is

$$\max_{z_t} E\{ \sum_{t=1}^{\infty} \alpha^{t-1} g(x_t, y_t) \} \qquad (15.2)$$

where $g(x_t, y_t)$ is the value function for a harvest of y_t at stock level x_t and α is the discount rate. The optimal harvesting strategy that maximizes (15.2) under (15.1) may lead to harvests that fluctuate greatly. To smooth out such fluctuations Mendelssohn adds a cost for these fluctuations and modifies the objective functional as

$$\max E\{ \sum_{t=1}^{\infty} \alpha^{t-1} g(x_t, y_t) - \lambda |x_t - y_t - z_{t-1}| \}$$

where λ is a parameter measuring the cost of fluctuations in the harvest. As λ increases the optimal harvest policies tend to become smooth and converge towards a steady level of harvest.

The second example is from stochastic automata theory [36] which illustrates the learning mechanism of a stochastic automation under an unknown random environment which may be nonstationary. In the random environment $R(c_1, c_2, \ldots, c_r)$ with r states where c_i is the probability that it occupies the i-th state, the automation has to choose one. Each choice of ith state $(i=1,2,..,r)$ has probability $p_i(t)$ and a penalty either zero or one. A learning mechanism is a reinforcement scheme by which the probability vector $P(t) = (p_i(t))$ is changed to $P(t+1)$ by using information about the random environment in the form of inputs into and outputs from it. The objective of the stochastic automation is to find out the action which minimizes the average penalty

$$\min E\{ \sum_{i=1}^{r} p_i(t) c_i \} \qquad (15.1)$$

by using a learning or reinforcement scheme. A simple example may illustrate the economic implications of this stochastic choice problem. Assume that we have n players each planning to open a store in one of r regions $R_1, R_2, \ldots, R_r$. If player i chooses region R_j, his payoff ϕ_i is given by

$$\phi_i = a_j/n_j; \quad \begin{matrix} i=1,2,\ldots,n \\ j=1,2,\ldots,r \end{matrix} \qquad (15.2)$$

where n_j is the number of players choosing the same region R_j and $a_j = \bar{a}_j + \varepsilon_j$, when $\bar{a}_j$ is a constant and ε_j is a zero-mean random noise. We assume that each player can know only his own payoff in each round of the play with no information about the payoffs of other players, but he can observe output y_i and the response s from the environment which may affect his choice in the next period. The value of s which may be viewed as penalty (or reward) may belong to a closed set i.e., $0 \leq s \leq 1$. One could now introduce various learning schemes in this game with incomplete information. One such scheme considered by Baba [36] and utilized in his simulation experiments is of the following form:

$$p_i(t+1) = p_i(t) - k\theta s(1-p_i(t))\frac{h}{1-h} + \theta(1-s)(1-p_i(t))$$

$$p_j(t+1) = p_j(t) + k\theta s(1-p_j(t))\frac{h}{1-h} - \theta(1-s)(1-p_j(t))$$

$$0<\theta<1, \; h=\min(p_1(t), \ldots, p_r(t)) \qquad (15.3)$$

$$p_1(0) = \ldots = p_r(0) = 1/r, \; 0<k\theta<1, \; i,j=1,\ldots,n$$

Here θ and k are suitable positive constants used in updating the probabilities $p_i(t)$ of choice. To see how various values of s may be assigned the following scheme is suggested:

$$\text{if} \quad \phi_t \geq \phi_{max}, \text{ then } s = 0$$
$$\phi_t \leq \phi_{min}, \text{ then } s = 1 \qquad (15.4)$$
$$\phi_{min} < \phi_t < \phi_{max}, \text{ then } s = 1 - \frac{\phi_t - \phi_{min}}{\phi_{max} - \phi_{min}}$$

where ϕ_t is the combined payoff at time t. If there is a coordinator or teacher who monitors the game, he may assign the value of s as above. This would be very similar to the Cournot-type conjectural variations game we considered before when implicit cooperation was introduced through public lotteries. The simulation experiments on the model defined by (15.2), (15.3), (15.4) reported by Baba [36] show the following results:

(a) players with learning scheme reinforcement have better convergence properties and lesser fluctuations in their choice patters than those who have no learning.

(b) the number of steps needed to reach the steady state solution is considerably reduced under a learning scheme. This implies that the loss from lack of coordination of different players' strategies may be considerably reduced.

5. Renewable Resources Model under Uncertainty

Problems of optimal decision-making and control in uncertain resource systems have recently attracted considerable attention from economists and other ecological researchers. Renewable resources present a more interesting problem in optimal control than exhaustible resources, because harvesting of renewable natural resources such as fish should be properly regulated in order to preserve the continuity of supply at all times.

The literature on renewable and exhaustible resource is now quite large [32], and problems of uncertainty can enter at different levels e.g., the stock of resources, the extraction or harvest rates, the price and costs all may contain random elements. We consider here two examples of a renewable resource like fisheries where optimal harvesting rates are to be determined over time. In the first example due to Pindyck [37] the resource stock dynamics is subject to a stochastic process and the market price p is exogenous. The second formulation is due to Clark [38] where Cournot-Nash equilibrium solutions are considered under deterministic demand. The implications of a stochastic demand function would be analyzed in this game-theoretic framework.

In a deterministic model of renewable resources the growth of resource stock $\dot{x} = dx/dt$ is given by

$$\dot{x} = f(x) - q(t) \qquad (16.1)$$

where $f(x)$ is a given growth function specifying the natural growth of the resource stock like fish and $q(t)$ is the extraction or harvest rate. For example f(x) could be the classical logistic function

$$f(x) = \alpha x\left(1 - \frac{x}{k}\right) \qquad (16.2)$$

where α is the intrinsic growth rate ($\alpha > 0$) and k is the invironmental carrying capacity; also the harvest rate $q(t)$ could be given by

$$q(t) = q_0 e(t) x(t) \qquad (16.3)$$

where $e(t)$ is the rate of fishing effort and q_0 is a parameter denoting the initial harvest rate.

The optimal harvesting rate $q(t)$ is determined by maximizing the discounted profit $\pi(t)$ over an infinite horizon i.e.,

$$\max_{q(t)} J = \int_0^\infty [p - c(x)] q(t) e^{-rt} dt \qquad (16.4)$$

subject to (16.1) and (16.2). Here total harvesting or extraction cost is $c(x)q(t)$ with marginal cost $c(x)$ decreasing and strictly convex and price p is determined by a downward sloping demand curve $p(q)$ so that the profit function $\pi = p(q)q - c(x)q$ is strictly concave. Define the Hamiltonian as

$$H = e^{-rt}(p - c(x))q(t) + \hat{\lambda}(t)(f(x) - q(t)) \qquad (16.5)$$
$$\text{where } \hat{\lambda}(t) = \lambda(t)e^{-rt} = \text{adjoint variable}$$

The optimal trajectory satisfies the following conditions by Pontryagin's maximum principle

CHAPTER III

$$\partial H/\partial q = 0 \Rightarrow \lambda(t) = p - c(x) \qquad (16.6)$$

$$\hat{\lambda} = d\hat{\lambda}/dt = (\dot{\lambda} - \lambda r)e^{-rt} = -\partial H/\partial x \Rightarrow \dot{\lambda} = \lambda(r - f'(x)) + c'(x)q$$

$$\dot{x} = f(x) - q(t)$$

and

$$\lim_{t \to \infty} e^{-rt}\lambda(t) = 0 \quad \text{(transversality)}$$

Here prime denotes partial derivative with respect to the argument. At the steady state equilibrium $\dot{p} = \dot{x} = 0$ yielding the pair (p^*, x^*) at which it holds

$$p^* = c(x^*) - \frac{c'(x^*)f(x^*)}{r - f'(x^*)}$$

$$f(x^*) = q^*$$

The model above defines a competitive market equilibrium where the agents are price takers i.e., price p is exogenous in the maximization but p and q satisfy the total market demand condition. If the resource is extracted by a monopolist instead, then along the optimal trajectory equations of (16.6), the price variable is replaced by marginal revenue. The monopolist then under-harvests at the optimum, so that the resource stock is larger than under competition.

Now under uncertain stock dynamics equation (16.1) is replaced by the stochastic differential equation

$$dx = [f(x) - q(t)]dt + \sigma(x)dz \qquad (17.1)$$

with $\sigma'(x)>0$, $\sigma(0)=0$ and $dz = \varepsilon(t)\sqrt{dt}$. Here $\varepsilon(t)$ is assumed to be serially uncorrelated and normally distributed random variable with unit variance i.e., $z(t)$ is a Wiener process. Under competitive market assumptions each firm is a price-taker and maximizes the present value V_i of profits:

$$V_i = V_i(x) = \max_{q_i} E_t \int_t^\infty \pi_i(\tau) e^{-r(\tau-t)} d\tau \qquad (17.2)$$

where it is assumed that the firm is risk-netural, profit is $\pi_i = [p-c(x)]q_i$ and marginal cost is identical for each of n firms. By using the fundamental recursive principle of optimality of dynamic programming [37] one could write for the optimal trajectory:

$$rV_i - \max_{q_i} \{pq_i - c(x)q_i + [f(x) - nq_i]V_{i,x} + \tfrac{1}{2}\sigma^2(x)V_{i,xx}\} \qquad (17.3)$$

where subscripts other than i denote partial derivatives i.e., $V_{x,i} = \partial V_i/\partial x$. Since the equation is linear in q_i, we obtain for the optimal harvest rate for each producer

$$q_i = \begin{cases} q_{i,max}, & \text{if } p-c(x) > nV_{i,x} = \bar{V}_x \\ 0, & \text{if otherwise} \end{cases}$$

where $\bar{V} = nV_i$ is the aggregate value of resource stock to producers. With a downward sloping demand curve, market clearing requires that total output $q=nq_i$ satisfies the relation

$$\bar{V}_x = p(q) - c(x) \qquad (17.4)$$

If the rent $\bar{V}_x$ is known, then the optimal harvest rate q^* can be solved from (17.4) as

$$q^*(x) = p^{-1}(\bar{V}_x + c(x)) \qquad (17.5)$$

If $\bar{V}_x$ is unknown, we have to solve the functional equation derived from (17.3) as

$$r\bar{V} = \int_0^{q^*(x)} p(q)dq - c(x)q^*(x) + [f(x) - q^*(x)]\bar{V}_x + \tfrac{1}{2}\sigma^2(x)\bar{V}_{xx}$$

On substituting the equilibrium harvest rate $q^*(x)$ from (17.5) into the stock dynamics equation (17.1) the optimal trajectory can be derived as

$$dx = [f(x)-q^*(x)]dt + \sigma(x)dz$$

Also, one can derive along the optimal trajectory a dynamic condition analogous to the second equation of (16.6) as follows

$$\frac{(1/dt)E_t\, d(p-c)}{p-c} + f'(x) - \frac{c'(x)q^*}{p-c} = r + \sigma'(x)\sigma(x)A(x) \qquad (17.5)$$

where $A(x) = -\bar{V}_{xx}/\bar{V}_x$ may be thought of as an index of absolute risk aversion. The deterministic case is

$$(1/dt)\, d(p-c) = r(p-c) - f'(x)(p-c) + c'(x)q \qquad (17.6)$$

On comparing (17.5) with (17.6) Pindyck [38] has derived some important consequences of stochastic fluctuations: (i) since the growth function

$f(x)$ is assumed to be concave, stochastic fluctuations in x reduce the expected rate of growth of x, (ii) since stochastic fluctuations tend to reduce the value of the stock, there is an incentive to reduce the stock level by harvesting faster and (iii) fluctuations increase expected harvesting costs over time and hence creates an incentive to harvest at a faster rate.

Two comments may be made here about this formulation. One is that this does not incorporate the possibilities of several fishing cycles or harvest policies for varying levels of initial harvest rate q_0 corresponding to the given initial population level x_0. In a simulation study Chen and Ahmed [39] have determined optimal fishing cycles by requiring that the total market demand be satisfied and the harvest rate does not have large fluctuations. In the stochastic case the need for smoothing out large fluctuations in the harvest rates is much greater, since an unpredictable catastrophe like oil spill may drive the fishery into extinction.

A second point is that the dynamic equation of growth e.g., (16.1) in the deterministic case is in many cases an approximation for a harvesting rate with delay so that the actual dynamics is

$$\dot{x} = f(x(t)) - q(t-\tau)$$

where $\tau > 0$ is the delay. This type of mixed difference-differential equation tends to contribute to more fluctuations. This emphasizes the need to develop more robust harvest policies which need not be fully optimal but suboptimal and more smooth.

Next we consider the second example [38], where we have two players in a Cournot-Nash differential game in deterministic form, with fishing effort $e_i(t)$ used in harvesting is to be optimally determined. The objective of each player is to maximize the present value of long run profits i.e.,

$$\underset{e_i(t)}{\text{Max}} \; J_i = \int_0^\infty e^{-rt} \pi_i(x(t), e_i(t)) dt$$

where profit is

$$\pi_i = pq_i - c_i e_i = (pk_i x - c_i) e_i \qquad (18.1)$$

and the other conditions are:

$\dot{x} = f(x) - q_1(t) - q_2(t)$

$x(0) = x_0 > 0; \; 0 \leq e_i(t) \leq e_i^{max}; \; 0 < x(t) < \bar{x}$

$x(t) \geq 0; \; q_i(t) = k_i e_i(t) x(t)$

$e_i(t)$ = rate of fishing effort by player i

c_i = constant marginal cost of fishing effort

k_i = catchability coefficient

$p = p(t)$ = landed price of fish

$f(0) = f(\bar{x}) = 0$ and $f(x) > 0$ for $0 < x < x^{max}$, where x^{max} is the maximum carrying capacity.

r = exigenous positive discount rate.

The demand function is taken to be of a linear form

$$p = a - b(q_1 + q_2) \qquad (18.2)$$

A Cournot-Nash (CN) equilibrium in non-zero sum game is then given by a pair $e_1^*(t)$, $e_2^*(t)$ which satisfy the following conditions for all feasible trajectories $e_i(t)$, $0 \leq t < \infty$:

$$J_1(e_1^*(t), e_2^*(t)) \geq J_1(e_1(t), e_2^*(t)), \text{ for all } e_1(t)$$

(18.3)

$$J_2(e_1^*(t), e_2^*(t)) \geq J_2(e_1^*(t), e_2(t)), \text{ for all } e_2(t)$$

The optimal trajectory of the model (18.1) for each i=1,2, is characterized by Pontryagin's maximum principle as

$$\dot{\lambda}_i = \lambda_i [r - f'(x) + K] - e_i(ak_i - 2bKk_i x)$$

$$\dot{x} = f(x) - Kx; \quad K = \sum_{i=1}^{2} k_i e_i$$

$$ak_i x - c_i - bx^2(K+k_i)k_i - \lambda_i k_i x = 0$$

(18.4)

$$\lim_{t \to \infty} \lambda_i e^{-rt} = 0$$

Assuming a symmetric case i.e., $k_i = k_0$, $c_i = c_0$ and n players one could derive from (18.4) the following steady state equilibrium values $\bar{e}$, $\bar{q}$, $\bar{x}$

$$\bar{e} = \sum_{i=1}^{n} \bar{e}_i = \frac{n(ak_0 \bar{x} - c_0)}{(n+1)bk_0^2 \bar{x}^2} - \frac{\bar{\lambda}}{bk_0^2 \bar{x}^2 (n+1)}$$

$$\bar{q} = \sum_{i=1}^{n} \bar{q}_i = k_0 \bar{x} \bar{e}$$

(18.5)

$$f(\bar{x})/\bar{x} = k_0 \bar{e}$$

$$\bar{\lambda} = \frac{[a - 2bf(\bar{x})]f(\bar{x})}{\bar{x}(r + k_0 \bar{e} - f'(\bar{x}))}$$

If $f(x)$ is a logistic function i.e.,

$$f(x) = \alpha x(1 - \frac{x}{x^{max}})$$

then $\bar{x} = x^{max}(1 - \frac{k_0 \bar{e}}{\alpha}), \frac{k_0 \bar{e}}{\alpha} < 1$

also $\lim_{n \to \infty} \bar{e} = (ak_0 x - c_0)(bk_0^2 \bar{x}^2)$

$\lim_{n \to \infty} \bar{q} = \frac{a}{b} - \frac{c}{bk_0 \bar{x}}$

$\lim_{n \to \infty} \bar{p} = \frac{c}{k_0 \bar{x}^0}, \bar{x}^0 = \lim_{n \to \infty} \bar{x}$

It is clear that increasing n has the effect of increasing the total harvest at the equilibrium and hence decreasing the steady state stock level of the renewable resource; the effect of increasing the discount rate r is similar.

A simple way to introduce demand uncertainty in this model is to formulate the inverse demand function with an additive error:

$$p = a - b \sum_{i=1}^{n} q_i + \varepsilon$$

where ε has a zero mean, constant variance σ^2 and is independent of total demand quantity. The objective function is then revised to include a term $\frac{w}{2} q_i^2 \sigma^2$ for the risk aversion of each player:

$$\text{Max}_{e_i} J_i = \int_0^\infty e^{-rt}(\pi_i - \frac{w}{2} q_i^2 \sigma^2) dt$$

where var $\pi_i = q_i^2 \sigma^2$ = variance of profit w is a nonnegative weight.

The risk aversion term may also be interpreted as an attitude to smooth out fluctuations in profit. The higher the weight w or σ^2, the lower would be the fishing effort and the total harvest in equilibrium. It is clear that cooperative, Pareto and other minimax solutions could be easily incorporated in this framework. Some theoretical and empirical applications of related models are made by Sengupta [26]. It may be noted that these differential game models are very similar in their structural characteristics to market models of limit pricing.

CHAPTER III

References

1. Milgrom, P and J. Roberts. Limit pricing and entry under incomplete information. Econometrica 50(1982), 443-460

2. Dixit, A. The role of investment in entry deterrence. Economic Journal 90(1980), 95-106

3. Sengupta, J.K. Optimal control in limit pricing under uncertain entry, in Information and Efficiency in Economic Decision, Dordrecht: Martinus Nijhoff, 1985

4. Kreps, D.M. and A.M. Spence. Modelling the role of history in industry organization and competition, in Issues in Contemporary Microeconomics and Welfare, Albany: State University of New York Press, 1985

5. Kamien, M.I. and N.L. Schwartz. Conjectural variations. Canadian Journal of Economics 16(1983), 191-211

6. Fershtman, C. and M.I. Kamien. Conjectural equilibrium and strategy spaces in differential games, in G. Feichtinger ed. Optimal Control Theory and Economic Analysis, vol. 2, New York: Elsevier Science Publishers, 1985

7. Starr, A.W. and Y.C. Ho. Nonzero-sum differential games. Journal of Optimization Theory and Application 3(1969), 184-208

8. Bresnahan, T.F. Duopoly models with consistent conjectures. American Economic Review 71(1981), 934-945

9. Harsanyi, J.C. Solution concepts in game theory. International Journal of Game Theory 5(1975), 39-54

10. Aumann, R.J. Subjectivity and correlation in randomized strategies. Journal of Mathematical Economics 1(1974), 67-96

11. Moulin, H. Cooperation in mixed equilibrium. Mathematics of Operations Research 1(1976), 273-286

12. Sengupta, J.K. Constrained nonzero sum games with partially controllable strategies. Journal of Optimizaiton Theory and Applications 31(1980), 50-62

13. Restrepo, R.A. Games with a random move, in: M. Dresher et. al. eds. Advances in Game Theory. Princeton: Princeton University Press, 1964, 29-38

14. Tintner, G. and J.K. Sengupta. Stochastic Economics. New York: Academic Press, 1972

15. Bharucha-Reid, A.T. Elements of the Theory of Markov Processes and Their Applications. New York: McGraw Hill, 1960

16. Brock, W.A. On models of expectations that arise from maximizing behavior of economic agents over time. Journal of Economic Theory 5(1972), 348-376

17. Newbery, D.M.G. Commodity price stabilization in imperfect or cartelized markets. Econometrica 52(1984), 563-578

18. Goreux, L.M. Optimal Rule for Buffer Stock Intervention. Mimeographed Report, IMF Research Department DM/78/7, Washington, D.C., 1978

19. Nguyen, D.T. The effects of partial price stabilization on export earnings instability and level, in: Commodities, Finance and Trade: Issues in North-South Negotiations. Westport, Connecticut: Greenwood Press, 1980

20. Hughes-Hallett, A.J. Commodity Market Stabilization and North-South Income Transfers: An Empirical Investigation. Unpublished Paper, December 1984

21. Whittle, P. Risk-sensitive Linear quadratic Gaussian control. Advances in Applied Probability 13(1981), 764-777

22. Geer, T. An Oligopoly: The World Coffee Economy and Stabilization Schemes. New York: Dunellen, 1971

23. Reynolds, P.D. International Commodity Agreements and the Common Fund. New York: Praeger Publishers, 1978

24. Calmfors, L. Long run effects of short-run stabilization policy: an introduction to a symposium. Scandinavian Journal of Economics 84(1982), 133-146

25. Narendra, K.S. and R.V. Monopoli eds. Applications of Adaptive Control. New York: Academic Press, 1980

26. Sengupta, J.K. Information and Efficiency in Economic Decision. Dordrecht: Martinus Nijhoff Publishers, 1985

27. Taylor, J.B. Methods of efficient parameter estimation in control problems. Annuals of Economic and Social Measurement 5(1976), 339-347

28. Stenlund, H. On the estimation versus control problem in cautions control, in: Brannas, D. et. al. eds. Econometrics and Stochastic Control in Macroeconomic Planning. Stockholm: Almqvist and Wiksell, 1981

29. Clarke, D.W. Some implementation considerations of self-tuning controllers, in: F. Archetti and M. Cugiani eds. Numerical Techniques for Stochastic Systems. Amsterdam: North Holland, 1980

30. Keyser, R.M.C. and A.R. van Cauwenberghe. Self-tuning prediction and control. International Journal of Systems Science 14(1983), 147-168

31. Astrom, K.J. and B. Wittenmark. On Self-tuning regulators. Automatica 9(1973), 185-200

32. Clarke, D.W. and P.J. Gawthrop. Self-tuning controller. Proceedings of IEEE Society 122(1975), 929-934

33. Mangel, M. Decision and Control in Uncertain Resource Systems. New York: Academic Press, 1985

34. Gleit, A. Optimal harvesting in continuous time with stochastic growth. Mathematical Bioscience 41(1978), 111-123

35. Mendelssohn, R. A systematic approach to determining mean-variance trade-offs when managing randomly varying populations. Mathematical Bioscience 50(1980), 75-84

36. Baba, N. Use of learning automata for stochastic games with incomplete information. International Journal of Systems Science 17(1986), 129-140

37. Pindyck, R.S. Uncertainty in the theory of renewable resource markets. Review of Economic Studies 51(1984), 289-303

38. Clark, C. Restricted access to common property fishery resources: a game-theoretic analysis, in: P. Liu ed. Dynamic Optimization and Mathematical Economics. New York: Plenum Press, 1980

39. Chen, Y.C. and N.U. Ahmed. An application of optimal control theory in fisheries management. International Journal of Systems Science 14(1983), 453-462

CHAPTER IV

Efficient Diversification in Optimal Portfolio Theory

1. Introduction

Recent developments in portfolio theory have used two interrelated approaches to specify efficient diversification of investment funds. One is the capital asset pricing model (CAPM), where the random return $\tilde{R}_p$ on a portfolio with n securities is viewed as follows:

$$\tilde{R}_p = R_F + \beta_p(\tilde{R}_M - R_F) + u_p \qquad (1.1)$$

where

$$\beta_p = \sum_{j=1}^{n} \beta_j x_j, \quad u_p = \sum_{j=1}^{n} u_j x_j$$

$$\sigma_p^2 = \text{var}(\tilde{R}_p) = \beta_p^2 \sigma_M^2 + \sum_{i=1}^{n} \sum_{j=1}^{n} x_i c_{ij} x_j$$

$$c_{ij} = \text{cov}(u_i, u_j) = E(u_i u_j)$$

Here x_j is the proportion of investment fund invested in security j, R_F is the risk-free rate of interest which may be assumed to be nonrandom, u_p is a random term with zero mean and covariances $C = (c_{ij})$ assumed to be independently distributed of x_j and β_j, and finally $\tilde{R}_M$ denotes the return on the market portfolio, such as Standard and Poor's 500 or, Dow Jones Index having mean $\bar{R}_M$ and variance σ_M^2. In the CAPM framework the term $\beta_p^2 \sigma_M^2$ specifies the <u>systematic risk</u> arising from fluctuations in the overall market average yield, while the second term in σ_p^2, i.e., $x'Cx = \sum\sum_{ij} x_i c_{ij} x_j$ characterizes the <u>unsystematic risk</u> due to

CHAPTER IV

specific securities selected in the portfolio by the individual investor. The diversification of a portfolio is a measure of the extent to which the portfolio risk σ_p^2 can be reduced through appropriate selection of the vector $x = (x_j)$ of allocation ratios. The greater the diversification of a portfolio, the less the unsystematic risk. Also it has been found empirically that the unsystematic risk converges at a decreasing rate to zero as more securities are randomly added to that portfolio. However, for a small investor who intends to save on transactions and other costs, the random selection of additional securities may in effect pose the problem of choosing between over and under diversification.

The second approach to optimal portfolio theory is to characterize the mean variance efficiency frontier for an investor who has to decide on the optimal allocation of vector x. The following nonlinear program is used for the characterization:

$$\text{Minimize } \sigma^2 = \sum_{i=1}^{n} \sum_{j=1}^{n} x_i v_{ij} x_j$$

subject to

$$\sum_{i=1}^{n} m_i x_i \geq c \qquad (1.2)$$

$$\sum_{i=1}^{n} x_i = 1, \quad x_i \geq 0$$

where $\tilde{y} = \sum_{i=1}^{n} \tilde{r}_i x_i$ is the random return on the portfolio, with $\tilde{r}_i$ representing the return on security $i=1,2,\ldots,n$ assumed to be distributed with mean m_i and a covariance matrix $V = (v_{ij})$. Given the parameter vector $\theta = (m, V, c)$ one solves for the optimal allocation vector x* and the associated minimal portfolio risk σ_*^2 say. By varying c within a nonnegative domain, one can characterize the efficiency frontier $\sigma_*^2 = f(c)$ by a function f(c). Recently Szego [1] developed an optimal portfolio theory along this line, except that he specified the constraint set for the allocation vector x as follows:

$$R = \{x | m'x = c, x'e = 1\} \qquad (1.3)$$

where m and x are column vectors with n elements, prime denotes transpose and e is a column vector with each element unity. Note that short sales are permitted (i.e., $x_j < 0$) and expected return equals a positive number c. Any allocation vector x belonging to R is called admissible and the boundary B^n of the region of admissible portfolios in the plane (σ_*^2, c) is described by the parabola $\sigma_*^2 = f(c)$ where

$$f(c) = (\alpha\gamma - \beta^2)^{-1} (\gamma c^2 - 2\beta c + \alpha) \qquad (1.4)$$

with $\qquad \alpha = m'V^{-1}m, \ \beta = m'V^{-1}e, \ \gamma = e'V^{-1}e$

provided the covariance matrix V is nonsingular. The optimal allocation vector x* in the boundary B^n can be expressed as

$$x* = x*(c) = (\alpha\gamma - \beta^2)^{-1} \{(\gamma c - \beta) V^{-1}m + (\alpha - \beta c) V^{-1}e\} \qquad (1.5)$$

which is unique for any given c, provided the term $(\alpha\gamma - \beta^2)$ is not equal to zero. In case the covariance matrix V is singular,

CHAPTER IV

the n returns $\tilde{r}_i$ are not statistically independent and hence we have to consider either portfolios with fewer securities or, replace the inverse of V by pseudo-inverse. The pseudo-inverse is not unique however.

The efficiency frontier equations (1.4) and (1.5) are used by Szego to define correlated portfolios and a minimum variance portfolio. The covariance of any two portfolios with allocation vectors x(1), x(2) with respective expected returns c_1 and c_2 but the same variance-covariance matrix V is defined by

$$\text{cov}(x(1), x(2)) = (\alpha\gamma - \beta^2)^{-1} (\gamma c_1 c_2 - \beta(c_1 + c_2) + \alpha) \qquad (1.6)$$

Hence the correlation is

$$\rho(x(1), x(2)) = \text{cov}(x(1), x(2)/(\sigma_*(c_1) \sigma_*(c_2)). \qquad (1.7)$$

where

$$\sigma_*(c_1) = [(\gamma c_1^2 - 2\beta c_1 + \alpha)/(\alpha\gamma - \beta^2)]^{\frac{1}{2}}$$
$$\sigma_*(c_2) = [(\gamma c_2^2 - 2\beta c_2 + \alpha)/(\alpha\gamma - \beta^2)]^{\frac{1}{2}}$$

Any two portfolio vectors x(1) and x(2) are said to be V-orthogonal, if the covariance term defined in (1.6) is zero. This implies that

$$c_1 = (\gamma c_2 - \beta)^{-1} (\beta c_2 - \alpha)$$
and
$$x(c_1) = (\gamma c_2 - \beta)^{-1} [V^{-1}(ec_2 - m)].$$

The minimum variance portfolio x_{MV} is obtained by minimizing the optimal variance σ_*^2 in (1.4) with respect to c. Since γ is a positive parameter, the function $\sigma_*^2 = f(c)$ is strictly convex in c so that there exists a value $c_0 = \beta/\gamma$ of c at which the optimal variance σ_*^2 is minimized. Let v_* denote this minimum value. It is easy to show that $v_* = 1/\gamma$ and $x_{MV} = x(c_0)$. Let D bet the domain $D = \{c | \underline{c} \leq c \leq \bar{c}\}$ where $(\underline{c}, \bar{c})$ are the lower and upper bounds. If c_0 belongs to D, then the minimum variance portfolio is achievable, otherwise not. It is easy to show by means of the covariance formula (1.6) that the covariance between each boundary portfolio $x = x(c)$ and the minimum variance portfolio $x_{MV} = x(c_0)$ is constant and given by the term $1/\gamma$.

The above discussion of optimal investment allocation in static portfolio theory is incomplete in several respects. For one thing, it ignores the problem of estimation risk, i.e., that the parameters $\theta = (m, V)$ are to be statistically estimated from observations before they can be used to compute optimal allocation vectors. This raises issues of robustness, e.g., (a) how sensitive is the optimal allocation vector to the asymmetry in the distribution of portfolio returns, and (b) how reliable is the optimal allocation vector in terms of the safety first criterion say, when the probability distribution of portfolio returns is either unknown or, known to belong to a nonnormal class.

A second set of issues is concerned with the problem of optimally revising a given portfolio, when new information becomes available, or new estimates of the parameters can be made. Three types of attempts have been made in the current literature for the portfolio revision problem. The first views the individual investor's revision problem as one of adjustment towards efficient market conditions, so that if the latter change, one has to readjust his portfolio. This approach ignores however the empirical evidence that the impact of information on individual investors is not perfect and the security-relative information market is not uniformly efficient either across securities or over time. A second type of approach due to Szego [1] considers theoretically the problem of enlarging the size of the existing portfolio and its impact on overall risk and return. However no systematic evaluation of the impact of enlarging a given portfolio has been attempted by this approach, except the case when enlargement leads to a singular variance-covariance matrix. The third approach due to Bawa, Brown and Klein [2] considers a Bayesian framework of revision. Thus, starting from a prior distribution of portfolio returns and combining it with sample information they determine the predictive distribution of portfolio returns. If this predictive distribution is normal or approximately so, then the mean-variance analysis still holds. However if the predictive distribtuion of portfolio returns is not normal, then the variance may prove to be a very incomplete measure of risk. In the latter case higher moments, e.g., skewness and kurtosis measures may have to be incorporated. As a matter of fact some empirical estimates of the influence of the skewness measures on security returns reported by Beedles [3] turn out to be significantly different from zero, thus implying some asymmetry in the distribution of returns.

A third set of issues is concerned with the comparison of alternative portfolios known as the mutual funds, which are available to the investors in the open capital market. When traded in the open market the mutual funds list their specific portfolio objectives into four groups, e.g., (a) growth fund emphasizing more on the growth of capital, (b) balanced fund emphasizing more risk aversion or risk reduction, (c) income fund with emphasis on income rather than growth, and (d) the growth-plus-income fund, which seek to combine the objectives of capital growth and regular income earnings. These four groups of mutual funds may be viewed as four risk classes with possibly unequal degrees of risk aversion as measured in the Arrow-Pratt sense. The existence of these risk classes has two implications for the optimal portfolio theory. First, the two objectives of growth and balanced diversification may have unequal impact on the two groups of investors and the existing portfolio theory may need to be modified and generalized. Second, each fund may be treated as a security and the investor may consider the problem of optimal mixing of different funds. This leads to a two-stage optimization problem - as in stratified sampling. Denote the overall return $\tilde{y} = \sum_{j=1}^{N} w_j \tilde{y}(j)$ as the weighted average ($w \geq 0$, $\Sigma w_j = 1$) of returns ($\tilde{y}(j)$) of mutual fund j, where $\tilde{y}(j) = \tilde{r}'(j) x(j)$ with $x(j)$ denoting an allocation vector with m_j

elements. Total variability of $\tilde{y}$ may then be decomposed into two parts: variability between the N funds (or strata) and variability within funds. Thus total variability may be minimized by choosing the weight vector $w = (w_1, w_2, \ldots, w_N)'$ and the allocation vector $x(j)$ in an optimal manner.

2. Estimation Risk in Portfolio Theory

One of the main reasons for testing the optimal allocation rules in portfolio choice for robustness is the presence of estimation risk associated with the statistical estimates of the mean variance parameters from sample observations. The asymmetry associated with ignoring such estimation risks has been strongly emphasized by Black and Scholes [4] who noted that the variances estimated on the basis of past data are subject to measurement error so that the spread in the distribution of estimated variances is larger than the true spread in the variances. Thus, the mean variance model using noisy estimates of variance tends to overprice options on high-variance securities and underprice options on low-variance securities. Further, the presence of estimation risk (or estimation error) may lead to singularity or near-singularity of the variance-covariance matrix. Thus suppose we need N observations to obtain a nonsingular estimate of the variance-covariance matrix V. We have however only N-1 actual observations. Then it is possible to find a portfolio with an estimated variance of zero, clearly an absurd result.

We may illustrate the important role of estimation risk in

optimal portfolio theory in three different ways, e.g., (a) showing the consequences of departures from the normal distribution, (b) using prior information as in Bayesian analysis and (c) utilizing the concept of a minimax portfolio.

Considering the first case of estimation risk, let $\bar{y} = N^{-1} \sum_{j=1}^{N} y_j$ and $s^2 = (N-1)^{-1} \sum_{j=1}^{N} (\tilde{y}_j - \bar{y})^2$ be the sample mean and sample variance of the portfolio return estimated from random sample of size N then for any infinite population, the variance or the squared standard error of s^2 is given by

$$\text{var}(s^2) = (N-1)^{-1} (2\sigma^4) + k_4 \qquad (2.1)$$

where

$$k_4 = \text{fourth cumulant} = E(\tilde{y} - \bar{y})^4 - 3\sigma^4$$
$$\sigma^2 = \text{population variance} = x'Vx$$

If the parent distribution is normal, then (2.1) reduces to

$$\text{var}(s^2) = (N-1)^{-1} (2\sigma^4) \qquad (2.2)$$

Hence it is clear that if we use sample mean returns $\bar{r}_i = \sum_{j=1}^{N} \tilde{r}_j/N$ to compute expected portfolio return $\bar{y} = \bar{r}'x$ where $\bar{r}$ is the sample mean returns vector, then the mean-variance portfolio model (1.3) reduces to

132   CHAPTER IV

$$\operatorname*{Min}_{x} \operatorname{var}(s^2), \; x \in R$$

where $R = \{x | \bar{r}'x = c, \; x'e = 1\}$ (2.3)

In the case of normal parent, this becomes

$$\operatorname*{Min}_{x \in R} f(x) = 2(N-1)^{-1} \sigma^2 (x'Vx)$$

Replacing V by its sample estimate S say, we obtain

$$\operatorname*{Min}_{x \in R} f(x) = 2(N-1)^{-1} (x'Sx)(x'Sx) \quad (2.4)$$

If the parent population is not normal, we obtain a different objective function as follows:

$$\operatorname*{Min}_{x \in R} f(x) = 2(N-1)^{-1} (x'Sx)^2 \left(1 + \frac{(N-1)G_2}{2N}\right), \quad (2.5)$$

where $G_2 = k_4/\sigma^4$ = Fisher's measure of kurtosis
$G_1 = E(\tilde{y} - \bar{y})^3/\sigma^3$ = skewness measure

We may note three important facts for this class of transformed portfolio models. First, the skewness in the original distribution measured by G_1 does not affect the stability of the sample variance s^2, i.e., the important factor which affects its stability is the fourth moment in the parent population. Second, even in the normal parent case, the objective function (2.4) is not quadratic in the allocation vector x; hence any x which minimizes x'Sx would not necessarily minimize the sample variance var(s^2) which is proportional to $(x'Sx)^2$. Thirdly, the last term in parenthesis in (2.5) shows the factor by which the variance of s^2 is inflated owing to nonnormality. Note that this factor is almost independent of N in the sense that as N → ∞,

$$[1 + \frac{(N-1)G_2}{2N}] \to [1 + \frac{G_2}{2}]$$

Hence the inflation factor may remain even with large samples.

Before we solve the optimization problem (2.4), we consider the subsidiary program:

$$\min_{x \in R} f(x) = 2h \cdot (x'Sx) \qquad (3.1)$$

where h is a positive parameter. Because x'Sx is positive definite by assumption, the objective function is strictly convex and it has only minimum for a given value of the parameter h. Now we have the following theorem:

Theorem 1

Suppose there exists an optimal solution (vector) x(h) of problem (3.1) such that it satisfies the condition

$$h = \hat{x}(h)' \, S \, \hat{x}(h) \qquad (3.2)$$

where $\hat{x}(h)$ is the optimal solution of (2.4). Then $x(h)$ is also an optimal solution of (2.4) and conversely.

Proof Using the Lagrangian function for problem (3.1)
$L = h \cdot (x'Sx) + 2\lambda(c - \bar{r}'x) + 2\theta(1 - x'e)$, one could write down the conditions to be satisfied by $\hat{x}(h)$ by following the Kuhn-Tucker theory:

$$\partial L/\partial x \text{ at } (\hat{x}, \hat{\lambda}, \hat{\theta}) = 4hS\hat{x} - 2\hat{\lambda}\bar{r} - 2\hat{\theta}e = 0$$
$$\partial L/\partial \lambda \text{ at } (\hat{x}, \hat{\lambda}, \hat{\theta}) = 2(c - \bar{r}'\hat{x}) = 0 \qquad (3.3)$$
$$\partial L/\partial \theta \text{ at } (\hat{x}, \hat{\lambda}, \hat{\theta}) = 2(1 - e'\hat{x}) = 0$$

For the objective function in (2.4), the solution vector $\hat{x}$ must satisfy the following necessary conditions

$$4(\hat{x}'S\hat{x}) S\hat{x} - 2\hat{\lambda}\bar{r} - 2\hat{\theta}e = 0$$
$$2(c - \bar{r}'\hat{x}) = 0; \; 2(1 - e'\hat{x}) = 0 \qquad (3.4)$$

The conditions (3.3) and (3.4 are completely equivalent. Hence if $\hat{x}(h)$ satisfies the condition (3.2), it must be the optimal solution of (2.4) also. Conversely, for any optimal solution $\hat{x}$ of (2.4) let us put $h = \hat{x}S\hat{x}$. Then $\hat{x}$ must also satisfy (3.3), since (3.3) and (3.4) have the equivalent structure.

Remark 1.1

This theorem is useful in developing a computational algorithm for solving the nonlinear program (2.4) - as follows:

Step 1: For $h = 1.0$ we solve the quadratic problem (3.1) and obtain an initial solution $\hat{x}(0)$ and therefore an initial value h_0 of h as $h_0 = \hat{x}(0)'S\hat{x}(0)$.

Step 2: Using the value of h_0 in place of h in (3.1), solve (3.1) to obtain the optimal solution $\hat{x}(1)$ and the associated value of h denoted by $h_1 = \hat{x}(1)'S\hat{x}(1)$. If h_1 satisfies (3.2), then $x(h_1)$ is an optimal solution of problem (2.4). If h_1 does not satisfy the condition (3.2), then we continue the iteration process. In case the covariance matrix S is semidefinite, one may have to check for degeneracy of any $\hat{x}(\bar{h})$ which also satisfy (3.2), i.e., $\bar{h} = \hat{x}(\bar{h})' S\hat{x}(\bar{h})$.

Theorem 2

Let $t(x) = 1 + (N-1) G_2/(2N)$ be such that for large N, the quantity $T(x) = (x'Sx) \frac{\partial t(x)}{t(x)\partial x}$ tends to zero for all $x \in R$. Suppose there exists an optimal solution $x(h)$ of problem (3.1), which also satisfies the condition

$$h = t(\hat{x}(h))(\hat{x}(h)' S\hat{x}(h)) \qquad (3.5)$$

where $\hat{x}(h)$ is the optimal solution of (2.5) in the asymptotic sense that $T(x)$ is negligible. Then $x(h)$ is also an optimal solution of the program (2.5) in the asymptotic sense.

Proof

By applying the necessary condition of optimality to the Lagrangian function

$$L = t(x)(x'Sx)^2 + 2\lambda(c - \bar{r}'x) + 2\theta(1 - e'x)$$

we obtain

$$\partial L/\partial x = t(\hat{x})(\hat{x}'S\hat{x})[4S\hat{x} + T(\hat{x})] - 2\hat{\lambda}\bar{r} - 2\hat{\theta}e = 0 \ .$$

This implies by the asymptotic condition:

$$4t(\hat{x})(\hat{x}'S\hat{x}) - 2\hat{\lambda}\bar{r} - 2\hat{\theta}e = 0 \qquad (3.6)$$

and also,

$$2(c - \bar{r}'\hat{x}) = 0; \ 2(1 - e'\hat{x}) = 0$$

An argument similar to that used in Theorem 1 then shows that any $\hat{x}$ satisfying (3.6) and (3.5) also solves the problem (2.5) in the asymptotic sense.

Remark 2.1

Since this result holds only in an asymptotic sense, the iterative method of computation may not be very convenient to apply. A more general procedure like the sequential unconstrained minimization technique (SUMT) may have to be applied.

2.1 <u>Prior</u> <u>Information</u> <u>in</u> <u>Portfolio</u> <u>Models</u>

Prior information on the mean variance parameters (m,V) of a portfolio is useful to an investor's decision problem in two ways. One is through updating the optimal allocation decision as more information becomes available. The second is in applying a minimax procedure to define a minimax portfolio in the estimation space if it exists. A minimax portfolio is one which minimizes the maximum risk defined suitably. By construction it is a robust portfolio, sine it characterizes the best of the worst decisions so to say.

Since the Bayesian method of updating in portfolio framework has been discussed by Bawa, Brown and Klein in some detail, we may discuss now how to characterize a robust decision in terms of the concept of a minimax portfolio. Consider first a simple case of a transformed portfolio where the returns vector is as follows:

$$W\tilde{r} = \tilde{q} \qquad (4.1)$$

Here W is a nonsingular matrix such that the covariance matrix V of $\tilde{r}$ is transformed as $D = WVW'$, where prime denotes transpose and D is a diagonal matrix with n diagonal elements formed out of the n eigenvalues of the positive definite matrix V. It is known that such a nonsingular matrix W always exists for any positive definite V. In terms of the new scale of measurement (4.1), any security i has return $\tilde{q}_i$ with mean μ_i and variance d_{ii} where d_{ii} is the diagonal element of the diagonal matrix D. Note that the covariance of returns of any two distinct securities is now zero. Hence the optimal portfolio model now minimizes $\sigma^2 = \sum_{i=1}^{n} d_{ii} x_i^2$ subject to $\Sigma \mu_i x_i = c$ and $\Sigma x_i = 1$. Suppose μ_i is unknown but $d_{ii} = \sigma_i^2$ say is known for each security

138   CHAPTER IV

$i = 1, 2, \ldots, n$ and we have to estimate a linear function $U = \sum_{i=1}^{n} \mu_i x_i$ where x_i are given numbers satisfying $\Sigma x_i = 1$. The loss function is assumed to be of the quadratic form

$$L(U, \delta) = (\delta - U)^2 \qquad (4.2)$$

where $\delta = \delta(q_{ij})$, is a decison function defined over the sample observations $\{q_{ij}\}$ with $i=1,2,\ldots,n$ and $j=1,2,\ldots,N$. Let Q denote the parameter space, i.e., a space of probability distributions on the sample space $\{q_{ij}\}$ indexed by $w \, \varepsilon \, Q$. The risk function $R(w, \delta)$ associated with the loss function (4.2) may then be written as

$$R(w, \delta) = E_w L(U, \delta) \qquad (4.3)$$

where E is expectation. If the decision fucntion δ is nonrandomized, then the subscript w in E has to be interpreted as fixed when the expectation is taken. For randomized decision functions one assumes that w is obtained by nature as a random variable with a probability distribution $p = p(w)$. Then a Bayes estimate with respect to the prior distribution p is an estimate δ which minimizes the average risk $\int R(w, \delta) \, dp(w)$. If the decision maker knew the prior distribution $p(w)$, he would choose this estimate as his best strategy. But in the absence of such knowledge of $p(w)$, he may use a minimax strategy as defined in two-person zero-sum game theory. A minimax estimate δ is one which minimizes the maximum risk defined by $\sup_{w \varepsilon Q} R(w,\delta)$. When minimax estimates exist, it has the robustness property of characterizing the best of the worst so to say. About the existence of minimax estimates there is a powerful theorem proved in Lehmann [5] as follows: Let p_N be a sequence of prior probability distributions, R_N be the sequence of associated Bayes risks; if R_N tends to R as $N \to \infty$ and if there exists some estimate δ for which $R(w,\delta) \leq R$, for all $w \varepsilon Q$, then δ is a nimimax estimate.

To apply the theorem in our case, assume that given $w \varepsilon Q$ the distribution of each security return is normal with an unknown mean μ_i and known variance $\sigma_i^2 = d_{ii}$, $i=1,2,\ldots,n$. Let μ_i be normally distributed as $N(0,\theta^2)$ with mean zero and unknown variance θ^2. We have to find Bayes solutions corresponding to the sequence $\{p_\theta\}$ of prior distributions of U resulting from the probability distribution of U. Let δ_θ denote a corresponding estimate of U. We have to investigate if the Bayes risks R_θ corresponding to δ_θ tend to a limiting value R when $\theta \to \infty$. If it does then any estimate which has its risk less than or equal to the limiting value R will be a minimax estimate by Lehmann's theorem.

It is apparent from the loss function (4.2) that the Bayes estimate and the Bayes risk R_θ are given respectively by the mean and the expectation of the variance of the conditional distribution of U given the sample $q = \{q_{ij}\}$. Note however that security sample means $\bar{q}_1, \bar{q}_2, \ldots, \ldots, \bar{q}_n$ are jointly sufficient for the parameters $\mu_1, \mu_2, \ldots, \mu_n$. Hence the conditional distribution of μ_i given $\bar{q}_1, \ldots, \bar{q}_n$ is normal

$N(M_i, V_i)$ with mean M_i and variance V_i where

$$M_i = (\sigma_i^2 + N\theta^2)^{-1} (N\theta^2 \bar{q}_i)$$
$$V_i = (\sigma_i^2 + N\theta^2)^{-1} (\theta^2 \sigma_i^2)$$

where N is the number of sample observations for each security return. Thus for given $(\bar{q}_1, \ldots, \bar{q}_n)$ the distribution of U is normal with mean $\sum_{i=1}^{n} M_i x_i$ and variance $\sum_{i=1}^{n} V_i x_i^2$. The Bayes estimate is $\delta_\theta(q) = \delta_\theta(\bar{q}_1, \ldots, \bar{q}_2) = \sum_{i=1}^{n} M_i x_i$ and the Bayes risk $R_\theta = \Sigma V_i x_i^2$. Now let us consider the limiting Bayes estimate

$$\lim_{\theta \to \infty} \delta_\theta(q) = \sum_{i=1}^{n} \bar{q}_i x_i = \delta^*(q) \text{ say} \qquad (4.5)$$

Since $\bar{q}_i$ is normally independently distributed as $N(\mu_i, \sigma_i^2/N)$, therefore $\Sigma \bar{q}_i x_i$ is normally distributed with mean $U = \Sigma M_i x_i$ and variance $\Sigma(\sigma_i^2 x_i^2/N)$. Hence the risk R for the minimax estimate $\delta^*(q) = \Sigma \bar{q}_i x_i$ is, for a fixed N

$$R = E(\sum_{i=1}^{n} \bar{q}_i x_i - U)^2 = E[\Sigma(\bar{q}_i - \mu_i) x_i]^2$$
$$\leq \Sigma \sigma_i^2 x_i^2 / N \qquad (4.6)$$

Thus the minimax estimate $\Sigma \bar{q}_i x_i$ is still a minimax estimate. Note that the minimax estimate is facilitated here by the existence of jointly sufficient statistics, i.e., sample means

$\bar{q}_1, \bar{q}_2, \ldots, \bar{q}_n$ are jointly sufficient for the parameters $\mu_1, \mu_2, \ldots, \mu_n$ in the normal distribution case. This property holds in Bayes theory for the special class of distributions known as the conjugate family.

To use the minimax estimate in optimal portfolio choice we may follow the same procedure as in (1.3) and (1.4) and derive

$$x_i^* = D \cdot \{(c \sum_i \bar{V}_i - \sum_i M_i \bar{V}_i) M_i \bar{V}_i$$

$$+ \bar{V}_i (\sum_i M_i^2 \bar{V}_i - c \sum_i M_i \bar{V}_i)\}$$

$$i = 1, 2, \ldots, n$$

$$\sigma_*^2 = D \cdot \{c^2 \sum_i \bar{V}_i - 2c \sum_i M_i \bar{V}_i + \sum_i M_i^2 \bar{V}_i\}$$

where
$$D = [(\sum_i \bar{V}_i)(\sum_i M_i^2 \bar{V}_i) - (\sum_i M_i \bar{V}_i)^2]^{-1}$$

$$\bar{V}_i = V_i^{-1}; \; M_i, \; V_i \text{ defined in (4.4)}$$

It is clear that as $\theta \to \infty$, each M_i tends to the sample mean return $\bar{q}_i$ and each V_i tends to σ_i^2/N but for finite θ there will be divergence between the pairs M_i and $\bar{q}_i$, or, V_i and σ_i^2/N.

We may now derive a minimax estimate in the general case when the transformation is not used. Let $\tilde{r}_1, \tilde{r}_2, \ldots, \tilde{r}_N$ be N random sample vectors from a multivariate normal distribution with an unknown value of the mean vector M and a specified precision matrix $\bar{V}$, where $\bar{V} = V^{-1}$ is the inverse of the variance-covariance matrix V. Although we assume V or, $\bar{V}$ to be known for the sake of simplicity, we have to approximate it for practical purposes by a suitable sample estimate. Let the prior distribution of M be a multivariate normal distribution $M \sim N(\mu, \bar{T})$ with mean vector μ and precision matrix $\bar{T} = T^{-1}$ such that μ is n-dimensional and $\bar{T}$ a symmetric positive definite matrix. Then the posterior distribution of M given r_j is a multivariate normal distribution with mean vector μ^* and precision matrix $(\bar{T} + N\bar{V})$, where

$$\mu^* = (\bar{T} + N\bar{V})^{-1} (\bar{T}\mu + N\bar{V}\bar{r}) \qquad (4.7)$$

$$\bar{r} = \sum_{j=1}^{N} \tilde{r}_j/N = \text{sample mean vector}$$

As the variance matrix T tends to infinite, $\bar{T}$ tends to zero and the limiting values of posterior mean and posterior variance become as follows:

$$\text{posterior mean} = \lim_{\bar{T} \to 0} \mu^* = [\lim_{\bar{T} \to 0} (\bar{T} + N\bar{V})]^{-1} [\lim_{\bar{T} \to 0} (\bar{T}\mu + N\bar{V}\bar{r})]$$

$$= \bar{r}$$

$$\text{posterior variance} = [\lim_{\bar{T} \to 0} (\bar{T} + N\bar{V})]^{-1} = (\bar{V})^{-1}/N = V/N$$

Thus we obtain the previous result that $\bar{r}'x = \sum_{i=1}^{n} \bar{r}_i x_i$ is a minimax estimate. It is clear therefore that replacing the two parameters (m, V) in the model (1.3) and (1.4) by $(\bar{r}, V/N)$ would define a minimax procedure. Cost of sampling or obtaining more observations may be easily introduced by a cost function

$$C = C(N) = aN + b(x'Vx/N)$$

where an optimal sample size N* can be determined by minimizing it, i.e.,

$$N^* = [b(x'Vx)/a]^{\frac{1}{2}}$$

3. <u>Minimax Portfolio</u>

The minimax estimate introduced in (4.5) and (4.7), which allows randomized decision rules for the decision-maker is not sufficiently general in character, since it is dependent on specific forms of prior distributions, e.g., conjugate family of probability distributions. We have to specify therefore a more generalized allocation procedure having a minimax property. To motivate this procedure we consider the notion of a limited diversification portfolio (LDP) introduce by Jacob [6]. Such a portfolio is most suitable for a small investor who chooses k our of n securities (k < n) to balance the two extremes of over versus under diversification. Empirical evidence on the benefits of diversification as a function of the number of securities held in the portfolio has been based on random selection. But random selections may not be optimal if the transaction costs and taxes prohibit frequent changes in investment allocation decisions by the small investor. On the basis of the CAPM framework (1.1), Jacob formulated a mixed integer quadratic programming model for a simplified LDP by assuming equal investment of funds among exactly k securities as follows:

$$\text{Min } \sigma_p^2 = ((1/k) \sum_{j=1}^{n} \beta_j x_j)^2 \sigma_M^2 + (1/k)^2 \sum_{j=1}^{n} x_j^2 \sigma^2(u_j)$$

$$\text{s.t. } (1/k) \sum_{j=1}^{n} x_j E(\tilde{R}_j) \geq c$$

$$\sum_{j=1}^{n} x_j = 1.0 \qquad (5.1)$$

$$\sum_{j=1}^{n} \delta_j \leq k; \; \delta_j - x_j \geq 0; \; x_j \geq 0, \; j=1,\ldots,n$$

$$\delta_j = 0 \text{ or, } 1.0 \text{ for all } j=1,2,\ldots,n.$$

Here k is an integer (k < n) denoting the size of the LDP and δ_j is a dummy variable which limits the size of the LDP. Some empirical simulations of this LDP model by Jacob showed that for a fixed k, $8 \leq k \leq 10$, the optimal LDP determined by (5.1) had variances uniformly lower than any of the randomly selected portfolios constructed for comparison purposes. To some extent these improvements in terms of lower variances of portfolio returns of LDP represent more efficient use of information concerning the market model parameters for those securities. For random selection there is no such efficient method of information utilization. One basic difficulty with this model (5.1) however is that it is a highly nonlinear program and only approximate solutions are obtained by Jacob for comparing with randomly selected allocation vectors. A more direct approach is to define a two-stage procedure following the minimax principle. In the first stage the investor selects k out of n securities (k < n) by maximizing the portfolio risk measured by the variance $\sigma^2(s) = s'(s) V(s) x(s)$ where $s=1,2,\ldots,K$ and K is the total number of selections each containing k securities. Let $v(s) = \max_{x(s)} \sigma^2(s)$ denote the maximum variance for each selection s, $s = 1,2,\ldots,K$. In the second stage he chooses one out of K selections by minimizing $v(s)$. Thus if s^0 is the second stage optimal choice then

$$v(s^0) = \min_{s \varepsilon I} v(s) = \min_{s \varepsilon I} \max_{x(s)} \sigma^2(s) \qquad (5.2)$$

where $I = \{1,2,\ldots,K\}$ is the index set of K selections. Since $\sigma^2(s)$ is a convex function of $x(s)$ for every s in the closed convex set defined by $R(s) = \{x(s) | m(s)' x(s) = c, x(s)'e = 1\}$, the maximum can occur only at an extreme point x^0 say, which is not a diversified portfolio. Thus for two securities x^0 would be of the form (1,0) or, (0,1). Hence we have to modify the constraint set for the allocation vector $x(s)$. To retain the diversified character of an allocation vector even when it maximizes the portfolio risk for a given selection s, we reformulate the constraint set as

$$R(s) = \{x(s) | (x(s) - x_0)' T(x(s) - x_0) \le 1\} \qquad (5.3)$$

where $T = \text{diag}(t_1, t_2, \ldots, t_k)$ is a diagonal matrix with positive diagonal elements denoting the weight placed on each security on the basis of its mean return and x_0 is the goal or the center of the ellipsoid defined by R(s) in (5.3). The convex set of constraints (5.3) may be viewed in two different ways. One is that the investor does not have any prior information on the allocation vector, so that x_0 is zero and T is an identity matrix. In this case the convex set specifies a restriction on the length of the allocation vector. A second interpretation is that the investor may have, in case of LDP models in particular, a special type of prior information in the form of a component by component restriction on x(s):

$$a_i \leq x_i(s) \leq b_i, \quad i=1,2,\ldots,k \qquad (5.4)$$

where (a_i, b_i) are finite lower and upper limits specified in advance for all selections. Prior specifications of the limits a_i, b_i by the decision-maker may imply a cautious and therefore risk-averse attitude of unwillingness to deviate too far. On writing these inequalities (5.4) as the cuboid:

$$\frac{|x_i(s) - (a_i + b_i)/2|}{(b_i - a_i)/2} \leq 1, \quad i=1,2,\ldots,k \qquad (5.5)$$

We may easily construct an ellipsoid defined by $(x(s) - x_0)' T(x(s) - x_0) = 1$ which contains the cuboid (5.5) and also fulfills the following conditions:

(i) the ellipsoid has the same center x_0 as the cuboid with a typical element $(a_i + b_i)/2$

(ii) the matrix T of the ellipsoid is diagonal with a typical diagonal element $t_i = 4/(k(b_i - a_i)^2)$

(iii) the ellipsoid has the minimal volume where volume is defined as $V_0 = \pi \prod_{i=1}^{k} (t_i^{-1})$.

and

(iv) the surface of the ellipsoid contains all corner points of the cuboid.

Since condition (iv) means that

$$\sum_{i=1}^{k} (\frac{a_i - b_i}{2})^2 t_i = 1 \qquad (5.6)$$

minimizing the volume $V_0 = \pi \prod_{i=1}^{k} t_i^{-1}$ subject to the condition (5.6) leads to the optimal values of t_i as $t_i = 4/(n(b_i - a_i)^2)$, $i=1,2,\ldots,k$. This defines the diagonal matrix $T = \text{diag}(t_1, t_2, \ldots, t_n)$ and the original prior information on $x_i(s)$ contained in the cuboid (5.5) is now reduced to the restriction $R(s)$ in (5.3) with x_0' being the row vector

$$(\frac{a_1 + b_1}{2}, \frac{a_2 + b_2}{2}, \ldots, \ldots, \frac{a_k + b_k}{2})$$

We thus obtain the maximum risk problem for each selection $s \in I$:

$$\max_{z(s) \in R(s)} \sigma^2(s) = z(s)' V(s) z(s) \qquad (5.7)$$

where

$$R(s) = \{z(s) | z(s)' \, Tz(s) = 1\}$$
$$z(s) = x(s) - x_0$$

Theorem 3

If the matrices $V(s)$ and T are symmetric and positive definite, then there always exists a minimax selection s^0 satisfying (5.2) under the restriction set (5.7). Associated with this minimax selection there exists a minimax portfolio $x(s^0)$ which is diversified. The minimax portfolio satisfies the generalized eigenvalue problem

$$[V(s) - \lambda(s) \, T] \, z(s) = 0 \qquad (5.8)$$

for $s = s^0$, where $\lambda(s)$ is the generalized eigengvalue.

Proof

On using $\lambda(s)$ as the Lagrange multiplier for the constraint $R(s)$ in (5.7) and applying the first order condition for a maximum, we obtain

$$(V(s) - \lambda(s) \, T) \, z(s) = 0$$

Since $V(s)$ and T are symmetric and positive definite, they are of rank n and hence there exist n positive values of $\lambda(s)$ which can be arranged as $\lambda_1(s) \geq \lambda_2(s) \geq \lambda_3(s) \geq \ldots \geq \lambda_k(s) > 0$. Since the index set I for $x \in I$ is finite, the minimum value of $\lambda_1(s)$, $s \in I$ exists. Hence if s^0 denotes the selection for which $\lambda_1(s^0) \leq \lambda_1(s)$, all $s \in I$ we have the minimax eigenvalue $\lambda_1(s^0)$

and its associated eigenvector $z(s^0)$ satisfying (5.8). But if x_0 is fixed by prior information, $z(s^0)$ leads to the allocation vector $x(s^0)$. Since the eigenvalue and the eigenvector are functions of all the elements of the matrices $V(s)$ and T, the minimax portfolio $x(s^0)$ is diversified, i.e., it cannot be an extreme point.

Remark 3.1

If the reference point x_0 is zero and T is the identity matrix I_k of order k, then $\lambda_1(s^0)$ is the minimax eigenvalue and $x(s^0)$ is the associated minimax portfolio. By minimizing over $\{\lambda_k(s), s \in I\}$ we can also determine a minimin eigenvalue say $\lambda_k(s_0)$ and the associated minimin portfolio $x(s_0)$.

Remark 3.2

Let $T = T_k$ and the security returns be all positively correlated, so that $V(s)$ has all positive elements and let it be indecomposable in the sense that there does not exist a nonsingular permutation matrix P such that

$$P^{-1}V(s)P = \begin{bmatrix} V_{11}(s) & V_{12}(s) \\ 0 & V_{22}(s) \end{bmatrix}$$

for every selection $s \in I$, where $V_{11}(s)$, $V_{12}(s)$ are square submatrices. Then by the Perron-Frobenius theorem, the minimax eigenvalue $\lambda_1(s^0)$ and the associated eignevector $x(s^0)$ which is the minimax portfolio are strictly positive such that the n elements of the vector $x(s^0)$ are all positive and hence they can be so normalized that they add up to unity.

Two implications of Theorem 3 on the minimax portfolio are most important from the investor's decision-making viewpoint. First, if the underlying probability distribution of security returns is very incompletely known, so that one has to replace $V(s)$ by its estimate $\tilde{V}(s)$, then it provides a cautious policy. Second, if $x_0 = 0$ and $T = I_k$ then there exist k positive eigenvalues for each selection $s \varepsilon I$:

$$[V(s) - \lambda(s) I_n] x(s) = 0$$

for each of which we can associate one eigenvector, i.e., one portfolio with k securities. These eigenvectors are all mutually orthogonal. Hence for the fixed selection $s^0 \varepsilon I$ where $x(s^0)$ is the minimax portfolio, we have $x^i(s^0)$ orthogonal to $x(s^0)$ where $x^i(s^0)$ is any eigenvector other than $x(s^0)$. For (k-1) eigenvalues other than $\lambda_1(s^0)$, we have (k-1) mutually orthogonal portfolios (i.e., eigenvectors). These portfolios characterize different degrees of diversification. For instance, if for a selection s^0, the difference between $\lambda_1(s^0)$ and $\lambda_k(s^0)$ is very small (i.e., the difference could be due to sampling fluctuations), then the minimax portfolio $x(s^0)$ may be said to be robust. Several implications of minimax policy towards portfolio choice have been analyzed by Sengupta [7], particularly when the posterior estimates of the variance-covariance matrix $V(s)$ may be obtained through additional sample observations.

Theorem 4

Let $T = I_k$ and $x_0 = 0$ and suppose that for every selection $s \in I$, the maximum eigenvalue $\lambda_1(s)$ and the associated eignevector $x^I(s)$ satisfy

$$m(s)' \, x^I(s) = c(s) > 0 \tag{6.1}$$

the condition of a positive average return $c(s)$, then the minimax variance $\lambda_1(s^0)$ is a nondecreasing convex function of the associated return $c(s^0)$. Hence for every minimax solection s^0, there exists some selection $s \in I$ for which $c(s) \geq c(s^0)$ and $\lambda_1(s) \geq \lambda_1(s^0)$.

Proof

Since the allocation vector $x^I(s)$ satisfies the condition (6.1), the maximiztation problem

$$\max_{x} x(s)' \, V(s) \, x(s) \tag{6.2}$$
$$\text{s.t.} \quad m(s)' \, x(s) = c(s)$$

would have the optimal solution vector $x(s) = x^I(s)$ for a fixed $c(s)$. Let $2r(s)$ be the Lagrange multiplier associated with the constraint of (6.2), then by applying the first order condition to the optimal Lagrangian function

$$L = x^I(s)' \, V(s) \, x^I(s) + 2r^*(s)(c(s) - m(s)' \, x^I(s))$$

we obtain

$$x^I(s)' \, (\partial L / \partial x^I(s)) = 0$$

which yields

$$r^*(s) = c(s)(m(s)' V^{-1}_{(s)} m(s))^{-1}$$
$$\lambda_1(s) = x^I(s)' V(s) x^I(s) = c^2(s)(m(s)' V^{-1}(s) m(s))^{-1}$$

This shows that for each fixed x and the fixed parameter set (m(s), V(s)), the maximum risk $\lambda_1(s)$ is a convex nondecreasing function of c(s). But since s ε I is a selection from a discrete set I, $\lambda_1(s)$ may not be strictly convex in c(s). But for the selection s^0, $\lambda_1(s^0)$ is the minimax eigenvalue over s ε I and hence there must exist an s ε I with $\lambda_1(s) \geq \lambda_1(s^0)$ for some s. Let s vary within the discrete set I as $s_1, s_2, \ldots, s_K$ for which we can compute the maximum risk levels $\lambda_1(s_1), \lambda_1(s_2), \ldots,$ $\lambda_1(s_K)$ and the expected returns $c(s_1), c(s_2), \ldots, c(s_K)$. Thus $\lambda_1(s)$ traces out a convex curve for c(s) if the lattice points are joined together. This is the risk return efficiency frontier. Let ŝ be the selection along the efficiency frontier for which it holds

$$v(\hat{s}) = \max_{s \varepsilon I} v(s) = \max_{s \varepsilon I} \max_{x(s)} \sigma^2(s)$$

then $\lambda_1(\hat{s}) \geq \lambda_1(s^0)$ and $c(\hat{s}) \geq c(s^0)$. The latter follows by the convexity of the risk-return efficiency frontier.

Remark 4.1

Let s̄ be the minimin selection defined by $v(\bar{s}) = \min_{s \varepsilon I} \min_{x(s)}$ $\sigma^2(s) = x(s)' V(s) x(s)$ under $x'(s) x(s) = 1.0$. If the eigenvalues are all distinct for each selection s, then $v(\bar{s}) < v(s^0) < v(\hat{s})$.

Remark 4.2

The minimin portfolio $x(\bar{s})$ associated with the minimin eigenvalue $\lambda_k(\bar{s}) = x(\bar{s})' V(\bar{s}) x(\bar{s})$ has lower variance than that of the minimum variance portfolio x_{MV} having variance $1/\gamma$, where γ is defined in (1.4). Here the minimum variance portfolio x_{MV} has to be computed on the basis of k securities to make them comparable with $x(s)$. This result follows because there are more constraints on the choice of the minimum variance portfolio.

4. Decision Rules for Portfolio Revision

Revising a portfolio may be viewed either as enlarging the size of a given portfolio by adding new securities or allowing more information to improve the estimates of the mean variance parameters of returns. We now consider these two types of revising a portfolio in terms of the two basic models we have presented, e.g., the quadratic efficiency frontier model (1.4) and the minimax portfolio model (5.7).

We start by developing some methods of characterizing the conditions under which a given portfolio is robust so that revisions are unnecessary so long as the specified conditions hold.

These conditions arise very naturally when one has to revise a portfolio of k securities say, e.g., one has to add a new security chosen from outside the given set. Let the reference set or the given set be denoted by S_0 and the set of all other securities by S. Let S_1 be a proper subset of S, called the comparison set. Then, one may define robustness as the property of optimality of a portfolio in S_0 relative to the comparison set S_1 or S as the case may be.

Definition 1. A set of k securities P_k belonging to the reference set S_0 is robust relative to the comparison set S_1, if the optimality of P_k in S_0 is not altered (i.e. maintained) by any choice from S_1. This may be termed as relative robustness.

Definition 2.

If the size of the portfolio $P_k \, \varepsilon \, S_0$ defined by the positive number $k \geq 2$ is such that its optimality is not altered (i.e. maintained) by any choice from S, then it is absolutely robust. Thus in this case the subset S_1 coincides with the set S.

It is clear therefore that the concept of robustness as defined here is related first to the optimality of the initial portfolio P_k and next to the comparison sets S_1 and S which depend on the information available to the decision maker at the current time point.

Now consider the quadratic efficiency frontier model (1.4), (1.5) and assume that the portfolio $x^*(c)$ in (1.5) is of size k, i.e., it contains k securities and the investor considers adding new securities say j-th security from the comparison set S_1, i.e., $j \, \varepsilon \, S_1$. We assume for simplicity that such enlargement is mean preserving, i.e., the expected return remains the same at the level c defined in (1.3). Instead of adding a new security, we may have more information to improve our original sample estimates of the mean variance parameters. In this case j may indicate channels for improving the original estimates of parameters. In both these cases, if they are mean preserving we would have the original parameters α, β, γ of (1.4) changed to α_j, β_j and δ_j respectively. The efficiency frontier equation would change to

$$\sigma_*^2(j) = b_j(c) = \delta_j^{-1}(\alpha_j - 2\beta_j c + \gamma_j c^2) \tag{7.1}$$

with

$$\delta_j = \alpha_j \gamma_j - \beta_j^2$$

where δ would be assumed to be positive. Furthermore the minimum value of $\sigma_*^2(j)$ is attained at

$$c_* = \beta_j/\gamma_j \quad \text{and} \quad \min_c \sigma_*^2(j) = 1/\gamma_j \tag{7.2}$$

where β_j would be assumed to be positive as is most likely in most realistic situations. Suppose the set S_1 of new selections is such that for all $j \in S_1$ it holds

$$\sigma_*^2(j) > \sigma_*^2 \tag{7.3}$$

and

$$\min_c \sigma_*^2(j) > \min_c \sigma_*^2 \tag{7.4}$$

then the reference portfolio $P_k \in S_0$ is robust relative to S_1. In this case the reference portfolio uniformly dominates all other enlarged portfolios constructed from S_1 for any value of $c > c_*$. However the conditions (7.3) or (7.4) may not hold for all selections $j \in S_1$. This provides the incentive to enlarge (or, change the dimension or structure of) the initial portfolio.

Theorem 5

Let $\alpha_j > \alpha$ and $\gamma < \gamma_j$, then the two quadratic efficiency frontiers $\sigma_*^2(j) = f_j(c)$ and $\sigma_*^2 = f(c)$ intersect at least once and at most twice, if they are not coincident in any interval of c. Let (c_0, σ_0^2) be any such intersection point. Then there exists a neighborhood around c_0, throughout which it holds that $\sigma_*^2(j) < \sigma_*^2$.

Proof

Since the quadratic functions $f(c)$ and $f_j(c)$ are strictly convex, the minimal values $v_*(j) = \min_c \sigma_*^2(j)$ and $v_* = \min_c \sigma_*^2$ exist and satisfy the inequality $v_*(j) > v_*$ since $\gamma < \gamma_j$. Since we have $\alpha_j = \delta_j f_j(0) > \alpha = \delta f(0)$, where $\delta = \alpha\gamma - \beta^2 > 0$, the two efficiency curves $f(c)$ and $f(c)$ must intersect at least once. Since both curves are strictly convex in c and monotonically rising to the right of their minimum points, they can intersect at most twice, provided they are not coincident. If (c_0, σ_0^2) is an intersection point where the curve $f_j(c)$ intersects the other curve from above, then to the right of the intersection point we must have an interval for which it holds that $\sigma_*^2(j) < \sigma_*^2$. If there is a second intersection point where $f_j(c)$ intersects $f(c)$ from below, we consider an interval to the left of the intersection point. By continuity such an interval must exist for which it holds that $\sigma_*^2(j) < \sigma_*^2$. Hence the neighborhood cannot be empty.

Remark 5.1

Let c_0 be the abscissa of an intersection point, to the right of which it holds $\sigma_*^2(j) > \sigma_*^2$. Then for all $c > c_0$, the initial portfolio $P_k \in S_0$ is robust relative to S_1.

Remark 5.2

The two marginal risk curves $M_j(c) = \dfrac{\partial f_j(c)}{\partial c}$ and $M(c) = \dfrac{\partial f(c)}{\partial c}$ intersect at a positive value of c, if each of the ratios β_j/β and γ_j/γ exceeds or falls short of the ratio δ_j/δ. If the point of intersection of the two curves $M_j(c)$ and $M(c)$, denoted by c_1 is positive and $\gamma_j/\delta_j > \gamma/\delta$ then for all $c > c_1$, the risk curve $f_j(c)$ is stochastically dominated by that of $f(c)$ in the sense of $\sigma_*^2(j) > \sigma_*^2$.

To see the implications of estimation risk we may now redefine the two sets S_0 and S_1. Let $\hat{m}_t$ be the sample mean vector of t independent observations, assumed to be normally distributed as $N(m, V)$ where the variance-covariance matrix V may be assumed to be known in the first case. In the second case we consider V to be unknown also. Now we replace the parameter vector m in (1.3) by the sample mean $\hat{m}_t$ where the latter is distributed normally as $N(m, \frac{1}{t} V)$ and the objective function $\sigma^2 = x'Vx$ is replaced by $\sigma_t^2 = x'Vx/t$, where the covariance matrix V is assumed known. By conditions of the central limit theorem, the sample mean $\hat{m}_t$ tends to normality $N(m, V/t)$ for a large class of distributions; hence asymptotically we would have the minimal variance $\tilde{\sigma}_{*t}^2$ portfolio for all nonnormal distributions obeying the central limit theorem.

Thus,

$$\tilde{\sigma}^2_{*t} = \tilde{\delta}_t^{-1} [\tilde{\gamma}_t c^2 - 2\tilde{\beta}_t c + \tilde{\alpha}_t]$$

where (7.5)

$$\tilde{\alpha}_t = t\hat{m}_t' V^{-1} \hat{m}_t , \quad \tilde{\beta}_t = t\hat{m}_t' V^{-1} e$$

$$\tilde{\gamma}_t = te'V^{-1}e , \quad \tilde{\delta}_t = \tilde{\alpha}_t \tilde{\gamma}_t - \tilde{\beta}_t^2 > 0$$

Let $\tilde{\sigma}^2_*$ denote an alternative minimal variance portfolio, where a sample statistic other than $\hat{m}_t$ is used such that any of the two following conditions holds:

(a) $\tilde{\sigma}^2_*$ does not tend to zero as $t \to \infty$

(b) $\tilde{\sigma}^2_*$ tends to zero as $t \to \infty$ but at a slower rate than that of $\tilde{\sigma}^2_{*t}$

Then one could prove the following result by straightforward calculations.

Theorem 6

Let S_0 denote the reference set portfolio characterized by the minimal variance $\tilde{\sigma}^2_{*t}$, whereas S_1 contain the alternative portfolios $\tilde{\sigma}^2_*$. Then for all nonnormal distributions converging to normality by the assumed conditions of the central limit theorem, the reference portfolio S_0 dominates over the alternative S_1 asymptotically in the sense that $\tilde{\sigma}^2_{*t}/\tilde{\sigma}^2_* < 1$ as $t \to \infty$.

Remark 6.1

The asymptotic robustness of the portfolio $\tilde{\sigma}^2_{*t}$ over that of $\tilde{\sigma}^2_*$ may be approached in two stages. In the first stage the size of the portfolio k is enlarged till we find an optimal size k* say. For a given optimal size k*, the second stage allows for sample size increases till an optimal non-dominated portfolio is found within the given class.

Remark 6.2

The quadratic form $[t(\hat{m}_t - m)' V^{-1}(\hat{m}_t - m)]$, which under the normality assumption has a chi-square distribution with k degrees of freedom can be used to set up a simultaneous confidence region for $m'\hat{x}_*$, where $\hat{x}_*$ is the optimal solution corresponding to $\tilde{\sigma}^2_{*t}$.

Now we consider the second case when the covariance matrix V in (7.5) is unknown but the normality assumption still holds. We replace V by its maximum likelihood estimator $\hat{V}$ which is adjusted to be unbiased:

$$\hat{V} = (t-1)^{-1} \sum_{j=1}^{t} (\tilde{m}_j - \hat{m}_t)(\tilde{m}_j - \hat{m}_t)'$$

Thus the sample mean vector $\hat{m}_t$ is normally distributed as $N(m, V/t)$ and the sample covariance matrix $\hat{V}$ has a Wishart distribution $W((t-1)^{-1} V, t-1)$. Replacing V by $\hat{V}$ in (7.5), a result analogous to Theorem 6 can be stated. Also since the quadratic form $[t(\hat{m}_t - m)' \hat{V}^{-1}(\hat{m}_t - m)]$ has the distribution of Hotelling's T^2 statistic, a simultaneous confidence region for $m'\hat{x}_*$ can be set up. Thus, the probability is $(1-\alpha)$ that the true value of portfolio return $m'\hat{x}_*$ falls in the critical region R:

$$R: \{\hat{m}_t'\hat{x}_* \pm (kF_{k,t-k})^{\frac{1}{2}}(\frac{t-1}{t-k})^{\frac{1}{2}}(\frac{\hat{x}_*'V\hat{x}_*}{t})^{\frac{1}{2}}\}$$

Note however that we have to use $\hat{x}_*$ instead of the true optimal decision x_* corresponding to the true parameter vector m, since the latter is not known. But asymptotically $\hat{m}_t \to m$ and then $\hat{x}_* \to x_*$.

Theorem 7

Let the two quadratic efficiency frontier equations be written as

$$g_j(c) = \gamma_j c^2 - 2\beta_j c + (\alpha_j - \delta_j \sigma_*^2(j)) = A_j c^2 + B_j \cdot c + D_j = 0 \quad (8.1)$$

$$g(c) = \gamma c^2 - 2\beta c + (\alpha - \delta \sigma_*^2) = Ac^2 + Bc + D = 0 \quad (8.2)$$

where $A_j = \gamma_j$, $B_j = -2\beta_j$, $D_j = \alpha_j - \delta_j \sigma_*^2(j)$ and $A = \gamma$, $B = -2\beta$, $D = \alpha - \delta\sigma_*^2$ and both D and D_j are nonzero. They have at least one root $c = c_0$ in common, if their resultant $R(g_j(c), g(c))$ is zero. The resultant R is a polynomial in the coefficients of $g_j(c)$ and $g(c)$ and can be written as the determinant

$$R = \begin{vmatrix} A_j & B_j & D_j & 0 \\ 0 & A_j & B_j & D_j \\ A & B & D & 0 \\ 0 & A & B & D \end{vmatrix}$$

or, as $R = (A_j D - D_j A)^2 - (A_j B - B_j A)(B_j D - D_j B)$.

Proof

This result is proved in standard texts on higher algebra. If $g_j(c) = 0$ has the two roots q_1, q_2 and $g(c) = 0$ has the two roots p_1 and p_2 so that $g(c) = A(c - p_1)(c - p_2)$, then the resultant R is defined as

$$R = A^2 A_j^2 \prod_{i,j=1}^{2} (q_i - p_j) = A_j^2 g(q_1) g(q_2)$$

Since both D and D_j are not zero, the resultant can be zero if and only if some difference $(q_i - p_j)$ is zero.

Remark 7.1

If the resultant $R = R(g_j(c), g(c))$ is not zero, then the two quadratic efficiency frontier equations do not intersect, i.e., do not have a root in common. If the parameters of $g_j(c) = 0$, $g(c) = 0$ are statistical estimates, then the resultant R may not differ from zero in a statistically significant sense in many situations, whence a common root may exist. In such cases the results of Theorem 5 may be applicable in a statistical sense.

Theorem 8

Let the new portfolio $\sigma_*^2(j) = f_j(c)$ denote additional observations only, with all parameters α, β, δ reestimated on the basis of enlarged samples and assume that it intersects the original portfolio $\sigma^2 = f(c)$ at (c_0, σ_0^2), then c_0 must satisfy the quadratic equation

$$[(\gamma_j/\delta_j) - (\gamma/\delta)] c_0^2 - 2(\beta_j/\delta_j - \beta/\delta) c_0 + (\alpha_j/\delta_j - \alpha/\delta) = 0$$

It follows therefore that c_0 is a real root if the discriminant Δ is nonnegative:

$$\Delta = [4(\beta_j/\delta_j - \beta/\delta)^2 - 4(\gamma_j/\delta_j - \gamma/\delta)(\alpha_j/\delta_j - \alpha/\delta)]^{\frac{1}{2}}$$

and it is the only root if $\Delta = 0$.

Proof

Since there exists at least one intersection point $c = c_0$, we must have $f_j(c) - f(c) = 0$ at c_0, which leads to the quadratic equation above and the results follow.

Remark 8.1

If the enlarged sample information is such that $\alpha_j = \alpha$ and $\delta_j = \delta$, then there exists a positive value of c_0 if $\beta_j > \beta$ and $\gamma_j > \gamma$.

Remark 8.2

If the two quadratic efficiency frontiers do not intersect for any real and positive value of c and $f_j(0) > f(0)$ for $c = 0$, then for any positive c, the original efficiency frontier dominates over the other in the sense $\sigma_*^2 < \sigma_*^2(j)$.

EFFICIENT DIVERSIFICATION IN OPTIMAL PORTFOLIO THEORY

Next we restrict ourselves to the minimax portfolios, which have a robustness property as analyzed before. For this purpose we restrict ourselves to a subset of risk averters or a subset of portfolios, for each of which the conditions $\hat{m}'x = c$, $e'x = 1$ hold, where $\hat{m}$ is a suitable sample estimate e.g., the sample mean vector $\hat{m}_t$ defined before. When agents or portfolios are so restricted, new results with stronger robustness which differ significantly from previously published results can be derived. Such types of subsets of risk averters have been used by other researchers to derive stronger results of diversification of risky portfolios. Restricted sets of this type are useful in those cases where it may be inappropriate to require that the variance of expected return, var $(\hat{m}'x)$ is a very reliably estimated measure. If this variance is not very reliably estimated, several estimates like $\hat{m}$ may result in returns $\hat{m}'x$ which will be statistically indistinguishable from an alternative estimate $\hat{m}'x$ say, due to the large variance of errors in the estimate.

For this restricted subset of risk averters or portfolios, one may use the maximum of variance $x'\hat{V}x$ as the worst possibility for each allocation and then select the best of the worst possibility. Since maximal variance will be unbounded otherwise, we adjoin a normalization condition $x'x = 1$. Let S_0 be the reference portfolio with k securities and S_1 denote the set of all enlarged portfolios indexed by $j = 1, 2, \ldots, J$ where J is the total number of enlarged portfolios.

Theorem 9

Let λ_0^* be the maximum eigenvalue of the portfolio in

S_0 having the varance-covariance matrix $\hat{V}$ of order k, which is assumed to be positive definite; also let λ_j^* be the maximal eigenvalue for each enlarged portfolio $j \, \varepsilon \, S_1$ having positive definite covariance matrices. If for all $j \, \varepsilon \, S_1$ it holds

$$\lambda_0^* \leq \min_{j \varepsilon S_1} \lambda_j^* \qquad (9)$$

then the optimal portfolio in S_0 is robust relative to any $j \, \varepsilon \, S_1$.

Proof

Since the variance-covariance matrices of the reference portfolio are symmetric and positive definite, their eigenvalues are all real and positive. Hence the maximal eigenvalues λ^*, λ_j^* are all positive. By the assumed conditions of the problem we have $\lambda_0^* \leq \lambda_j^*$, all $j \, \varepsilon \, S_1$. Hence in terms of maximum risk, the portfolio in S_0 has a risk level equal to or lower than any other $j \, \varepsilon \, S_1$. Hence the result.

Remark 9.1

If there exists a selection $j \, \varepsilon \, S_1$ for which the robustness property fails to hold i.e. $\lambda_j^* < \lambda_0^*$, then the reference portfolio can always be improved upon through diversification.

Remark 9.2

Let S be the union of the sets S_0 and S_1 and let the order k denoting the size of the portfolio be fixed. Let there be a total of G selections, each of size k. If λ_g^* and x_g^* be the maximal eigenvalue and the associated eigenvector for any selection $g \varepsilon G$, $G \varepsilon S$, then there must exist a selection $\lambda_0^* = \min_{g \varepsilon G} \lambda_g^* \leq \lambda_g^*$ which is robust in the sense of minimax. If, in addition, we consider only a restricted subset of G, such that the variance-covariance matrices for each selection in this subset have strictly positive elements (which necessarily holds true if each j is a mutual fund), then by the Perron-Frobenius theorem, there exists a strictly positive eigenvector x_0^* associated with λ_0^*. Hence if x_0^* is normalized so that its elements sum to unity, we may obtain the expected return level $\hat{m}'x_0^* = c_0^*$ say. One could thus construct a risk-return curve for each selection, from the restricted subset of G.

Next we compare a minimax portfolio with the mean-variance model (7.1). For this minimax portfolio, we consider each j as a mutual fund, so that we have to choose a composite portfolio of k mutual funds out of n funds. The returns of mutual funds are all positively correlated, so that for any selection g out of a total of G selections ($G = \binom{k}{n}$) we have a maximal positive eigenvalue λ_g^* and the associated eigenvector x_g^* with positive elements, provided the associated variance-covariance matrix is indecomposable. Likewise a minimal positive eigenvalue λ_{g*} and its associated eigenvector x_{g*} which may not have all positive elements may be obtained. Let $\lambda_0^* = \min_{g \varepsilon G} \lambda_g^*$ and $\lambda_{0*} = \min_{g \varepsilon G} \lambda_{g*}$ be the minimax and minimin eigenvalues with corresponding eigenvectors x_0^* and x_{0*} respectively. Note that x_0^* can be so normalized that its elements sum to unity.

Now for each selection $g \in G$ one could derive the quadratic efficiency frontier by model (7.1):

$$\sigma_*^2(g) = \delta_g^{-1}(\alpha_g - 2\beta_g c + \gamma_g c^2) \qquad (10.1)$$

where

$$\delta_g = \alpha_g \gamma_g - \beta_g^2 > 0 \quad \text{(by assumption)}$$

Theorem 10

The minimal variance $\sigma_*^2(g)$ of the efficiency frontier model (10.1) is bounded as follows:

$$\lambda_{g*} \leq \sigma_*^2(g) \leq \lambda_g^* \qquad (10.2)$$

Hence for any fixed value of $\sigma_*^2(g)$ satisfying (10.2), one could construct a mixture $p\lambda_{g*} + (1-p)\lambda_g^*$ for $0 \leq p \leq 1$ such that for a suitable $p = p_0$ we have $p_0 \lambda_{g*} + (1-p_0) \lambda_*^*(g) = \sigma_*^2(g)$. In other words, for every level of minimal variance generated by the quadratic efficiency frontier model (10.1), we could find a mixed portfolio on the basis of two extreme eigenvalues.

Proof

For any selection $g \in G$ the two extreme eigenvalues are obtained from maximizing and minimizing the variance $x_g' V_g x_g$ subject to $x_g' x_g = 1.0$. Since the eigenvector x_g^* has positive elements by Frobenius-Perron theorem, it can be normalized so that it sums to unity. The mean-variance efficiency model (7.1) has an identical requirement, i.e., $\Sigma x_i = 1$. But it has an additional constraint that $\Sigma m_i x_i = c$. Hence the bounds (10.2) hold good, since the minimum (maximum) of a constrained model cannot be lower (greater) than that of an unconstrained one. Let $\bar{\lambda} = p\lambda_{g*} + (1-p) \lambda_g^*$ be the linear convex combination of λ_{g*} and λ_g^* for $0 \leq p \leq 1$. Since $\bar{\lambda}$ is continuous in p and $\sigma_*^2(g)$ satisfies the bounds (10.2), there must exist a value p_0 of p at which $\bar{\lambda} = \bar{\lambda}(p_0) = \sigma_*^2(g)$. Note that the mixed strategy associated with $\bar{\lambda}$ can be defined as $\bar{x} = p_0 x_{g*} + (1-p) x_g^*$.

Remark 10.1

Suppose σ_*^2 of the mean-variance efficency model (7.1) satisfies for all $g \ \varepsilon \ G$

$$\sigma_*^2 \leq \lambda_g^* \text{ and } \sigma_*^2 \geq \lambda_{g*}$$

where σ_*^2 denotes the minimal variance portfolio of order $k + 1$ and both λ_g^* and λ_{g*} involve portfolios of order k. Then σ_*^2 is bounded by the maximin and minimax portfolios

$$\max_{g \varepsilon G} \lambda_{g*} \leq \sigma_*^2 \leq \min_{g \varepsilon G} \lambda_g^* = \lambda_0^*$$

Remark 10.2

If we obtain a maximum variance portfolio in terms of the mean variance model (7.1) by maximizing instead of minimizing the total variance for any selection $g \ \varepsilon \ G$, then the maximal variance $\sigma^{2*}(g)$ must satisfy the inequality $\sigma^{2*}(g) \leq \lambda_g^*$ and the associated allocation vector $x^*(g)$ must be less diversified than the vector x_g^* associated with λ_g^*. This follows because $x'Vx$ is a strictly convex function of x and hence its maximum can be attained only at an extreme point, where x satisfies $\Sigma x_i = 1$ in the model (7.1). If $k = 2$, this means x can be either $(0,1)$ or $(1,0)$. On the other hand for the eigenvalue problem, the vector x being an eigenvector depends through the eigenvalue on all the elements of the variance-covariance matrix.

In order to circumvent this difficulty we may define a minimax allocation in terms of the mean variance model (7.1) slightly differently. Since the partial derivative of portfolio risk $\sigma^2 = x'Vx$ with respect to allocation x may be viewed as marginal risk, i.e., $\partial\sigma^2/\partial x = 2Vx$ we may set up the following minimax allocation model:

$$\text{Min}_x \ u \text{ subject to}$$

$$2Vx \leq ue$$
$$m'x \geq c \qquad (10.3)$$
$$e'x = 1, \ x \geq 0$$

where u is a scalar, e is a vector with unit elements and the variance-covariance matrix V is such that Vx is positive. Note that this is a linear programming (LP) model, where the optimal vector x will still be diversified. If (x^0, u^0) solves the above LP model, then it follows

$$x^{0'}Vx^0 = u^0/2$$

5. Dynamic Rules of Revision

Rules of revisions take a dynamic character when new observations become available over time. The set of new observations and information may be viewed in two ways: the passive way when it is used to revise the original estimates and make them more reliable, and the active way when intertemporal decisions are considered with a planning horizon of several periods.

In the first case of revisions, we assume that more observations enable us to screen the data set by eliminating outliers or data points deviating very far from their mean levels so to say. Thus if the variance covariance matrix estimate $\hat{V}$ and the mean return vector $\hat{m}$ are trimmed so to say to $\hat{V}_t$ and $\hat{m}_t$, so that they are more robust, then in some sense the trimmed variance of portfolio return $y = \Sigma \tilde{r}_i x_i$ may be reduced by some suitable methods of trimming discussed, e.g., by Lehmann [5].

One may thus compare the untrimmed mean variance model with the trimmed one and estimate the relative gain from using the trimmed model. Several types of trimming have been discussed by Sengupta [8] in terms of truncation of the sampling distribution of observed returns. As an illustration consider portfolio returns (y) as a linear regression model

$$y = \mu + \varepsilon, \quad \mu = m'x \qquad (4.1)$$

where the errors ε are assumed to be independent random variables with zero mean and finite variance, and then introduce the constraint

$$m'x \geq c$$

for a preassigned level of $c > 0$. By redefining μ as $m'x-c$, the constraint implies for the regression model the restriction

$$y \geq 0$$

which implies that $\varepsilon \geq 0$ for the error. Thus if the error ε is normally distributed, the constrained error would not be so. The impact of such constraints on errors is two-fold. First, the efficiency of the conventional regression estimate $\hat{m}$ of m given x, derived from the unconstrained model would be considerably reduced for some c, since the probability $Prob(\hat{m}' x \geq c)$ may not be high. Recent econometric work on the estimation of efficient production functions has shown this very clearly. Second, the mean-variance properties of the truncated model would be radically different from those in the original model, whenever the truncation point c is not trivial. Also, some distributions are more sensitive to truncation and some are not and in the former case the mean-variance parameters are highly prone to outliers, as it has been shown by Neyman and Scott [9].

Let the truncated model be defined by $y^* = \max\{0,y\}$, where $y = m'x-c$ for a positive level of c preassigned by the decision-maker. These truncated variables y^* have also been called limited dependent variable models in econometric literature [10], where two approaches have been adopted for estimation. One is due to Amemiya [11] who assumed ε to be truncated normal and produced nonlinear maximum likelihood estimates of the parameters μ, σ^2 of the distribution of y. A second approach assumes that the truncated variable y^* has a skewed distribution and then it is transformed to approximate normality.

Our approach here is not one of estimation, although we consider such estimation problems to be very important in portfolio models. Two aspects of the truncation problem are emphasized here. The first deals with the mean variance relationships of the two variables y and y^* defined above, when the level of truncation c is varied along the efficiency frontier. Second, we would emphasize the element of additional risk produced when the sample mean estimates $\hat{\mu} = \hat{m}'x - c$ are required to satisfy the truncated model $\hat{y} \geq 0$. The first case deals with the specification bias, when the unconstrained model with y as the dependent variable is used instead of the truncated model with y^*. The assumption that the mean and variance parameters of the distribution of y are insensitive or neutral to any preassigned level of c is rejected by the truncated model, since the level of c critically affects the choice of an optimal allocation vector x. The second aspect emphasizes the point that in addition to the population variance $(x'Vx)$, there is an additional risk due to the variance: $Var(\hat{m}'x | \hat{m}'x \geq c)$, whenever the sample mean estimate $\hat{m}$ is used to satisfy the requirement $\hat{y} = \hat{m}'x - c \geq 0$ of the truncated model. Unless the additional risk tends to be small, at least under large sample conditions for the relevant range of values of c, the mean-variance efficiency frontier determined by any of the portfolio models considered before would tend to be non-robust and hence truncation-prone.

If y is normally distributed with mean $\mu = m'x - c$, $c > 0$ and variance $\sigma^2 = \text{var}(\varepsilon)$, then the mean and variance of the truncated variable y* can be computed as:

$$E(y^*) = \mu^* = \mu F(\frac{\mu}{\sigma}) + \sigma f(\frac{\mu}{\sigma})$$

$$\text{Var}(y^*) = \sigma^{*2} = \mu^*(\mu - \mu^*) + \sigma^2 F(\frac{\mu}{\sigma})$$

(11.1)

where F and f are the cumulative distribution and the probability density functions of a unit normal variate. Note that the variance in the truncated model is always less than that of the original model, i.e. $\sigma^{*2} < \sigma^2$ and as a function of the allocation vector x, it has the same general shape as $\mu = m'x - c$. Since $\mu^* > \mu$, it follows that $(\sigma^*/\mu^*) < (\sigma/\mu)$, i.e., the truncated model is more stable in terms of the coefficient of variation.

Theorem 11

If there exists an optimal solution $x_* = x_*(c)$ of the mean variance model given by (1.2) and (1.3), there must exist an optimal value c* of c which maximizes $h(c) = \mu^*(c) - \frac{r}{2} \sigma^{*2}(c)$, where a positive value of r denotes risk aversion. The strategy

EFFICIENT DIVERSIFICATION IN OPTIMAL PORTFOLIO THEORY

pair (x_*, c^*) has the property that given c at c^*, the allocation vector x_* minimizes the portfolio risk $x'Vx$, while given x at x_*, c^* maximizes the return from the appropriate truncated model.

Proof

Since both $\mu^*(c)$ and $\sigma^{*2}(c)$ are decreasing continuous functions of c within the domain $c_1 \leq c \leq c_2$, where c_1 and c_2 are any two fixed bounds, the curve $h(c)$ has a maximum for a fixed positive r; also the slopes of $(-\sigma^{*2}(c))$ and $\mu^*(c)$ are positive and negative respectively thus implying that they intersect at a common point. If c^* is the optimal value of c which maximizes $h(c)$, then on substitution, $x_*(c^*)$ specifies the optimal solution of the mean variance model given by (1.2) and (1.3).

Remark 11.1

Although the variance σ^{*2} of the constrained portfolio is always less than σ^2 for any finite μ and this variance σ^{*2} can be reduced by choosing c to be as large as possible, it may be difficult to realize this gain due to large errors involved in the process of estimation in a truncated world [8].

There is another way of looking at truncation, when one compares alternative portfolios rather than alternative securities. Thus let y_1 and y_2 be two distinct portfolios e.g., of a gven size k, where at least some securities are not common to each. Let $x(1)$, $x(2)$ denote the allocation vectors in these two portfolios, where an asterisk would denote optimal vectors. Depending on the bivariate probability distributions of y_1 and y_2, let us define the following sets of conditional means and variances:

$$E\{y_1|y_2\} = M_{12}; \ Var\{y_1|y_2\} = V_{12}$$

$$E\{y_1|y_2; \ y_1 \geq c, \ y_2 \geq c\} = M_{12}(c); \ c > 0$$

$$Var\{y_1|y_2; \ y_1 \geq c, \ y_2 \geq c\} = V_{12}(c); \ c > 0$$

(11.2)

$$E\{y_i|y_i \geq c\} = M_i(c), \ Var\{y_i|y_i \geq c\} = V_i(c), \ c > 0$$

It is clear that if there exists an allocation vector $x(2)$ in the alternative portfolio y_2 such that

$$M_{12} \geq M_1 \ \text{and} \ V_{12} \leq V_1 \qquad (11.3)$$

with at least one strict inequality

where M_i, V_i are the mean return and variance of portfolio y_i, then the given portfolio y_1 is not efficient against revision. Let S_1 be the comparison set of all portfolios like $y_2 \ \varepsilon \ S_1$, where the given portfolio y_1 is in reference set S_0. Then if the condition (11.3) holds, there exists at least one portfolio in S_1, which is more efficient than $y_1 \ \varepsilon \ S_0$. If for no $y_2 \ \varepsilon \ S_1$ the conditions (11.3) hold, but there exists a positive truncation level c such that

$$M_{12}(c) \geq M_1(c) \text{ and } V_{12}(c) \leq V_1 \qquad (11.4)$$

with at least one strict inequality

then the given portfolio is not efficient to truncation level c against revision. Thus a concept of c-efficiency can be defined as follows: A set of portfolios $y_1 \varepsilon S_0$ is c-efficient in a class of probability distributions F, against all other portfolios $y_2 \varepsilon S_1$, where S_1 is the appropriate comparison set generated by the probability distribution F and the truncation point c, if the conditions (11.4) fail to hold. If the c-efficiency property holds for all positive c in the relevant domain of the decision space, then $y_1 \varepsilon S_0$ is globally c-efficient; otherwise it is locally c-efficient.

Theorem 12

For any portfolio $y_i \varepsilon S_0$ which is normally distributed with mean M_i and variance V_i, there exists a positive c at which it holds

$$M_i(c) \geq M_i \text{ and } V_i(c) \leq V_i \qquad (12.1)$$

with at least one inequality strict. Furthermore for every pair of distinct portfolios $y_i \varepsilon S_0$, $y_j \varepsilon S_1$ which are normally distributed, there must exist a positive scalar c such that

$$0 \leq V_{ij}(c) \leq V_i$$

so that by appropriate truncation the conditional variance of $y_i \varepsilon S_0$ can be reduced as $V_{ij}(c) < V_i$.

Proof

The truncated means $M_i(c)$ and variances $V_i(c)$ can be directly calculated from the normal distribution of y_i with mean M_i and standard deviation $\sigma_i = V_i^{\frac{1}{2}}$ as follows:

$$M_i(c) = [1 - F(\frac{c-M_i}{\sigma_i})]^{-1} [M_i + \sigma_i f(\frac{c-M_i}{\sigma_i})]$$

$$V_i(c) = (M_i^2 + \sigma_i^2) + (1-F)^{-1} f(\frac{c-M_i}{\sigma_i})\{2M_i\sigma_i + c\sigma_i - M_i\} - \{M_i(c)\}^2$$

where $F = F(\frac{c-M_i}{\sigma_i})$ is the cumulative distribution function of a unit normal variate with density $f = f(\frac{c-M_i}{\sigma_i})$. Since $F(q)$ and $f(q)$ are positive fractions for any finite q, $M_i(c) \geq M_i$ for c; also for positive c, $V_i(c)$ is a monotonically declining function of $c > 0$, implying $V_i(c) \leq V_i(0) = V_i$. Furthermore, we have the conditional variance $V_{ij} = \text{Var}\{y_i | y_j\} = \sigma_i^2(1 - \rho^2)$ where ρ^2 is the squared correlation coefficient. Since $0 \leq \rho^2 \leq 1$, therefore we have

$$0 \leq V_{ij} \leq \sigma_i^2$$

But $\sigma_i^2(c) \leq \sigma_i^2$ for a positive truncation level c, so that

the truncated variance $\sigma_i^2(c)$ can be reduced below $\sigma_i^2 = V_i$.
Hence the result.

Remark 12.1

Although by truncation, improved portfolios satisfying the conditions (12.1) can be obtained if the normality assumption holds, the sample estimators for $M_i(c)$ and $V_i(c)$ having good properties like unbiasedness or minimum variance are very difficult to derive, due to nonnormality and hence to nonlinear maximum likelihood equations. For instance, the sample mean estimator $\bar{M}_i(c)$ based on t observations has the squared standard error

$$\text{Var}(\bar{M}_i(c)) = \frac{\sigma_i^2}{t} \delta(c) \qquad (12.2)$$

for large $t \to \infty$, where the factor $\delta(c)$ depends on the truncation point c. If $c = -\infty$, the degree of truncation is zero, the factor $\delta(c)$ converges to one; but for $c \to \infty$, the degree of truncation is 100% and $\delta(c)$ tends to infinity. Thus, with increase of c, the standard error of the sample mean estimator $\bar{M}_i(c)$ increases very fast. Thus, an optimal trade-off exists on the choice of c.

Remark 12.2

In case of positive truncation points c satisfying (12.1), one may be tempted to choose a level of c so high that $V_i(c)$ tends to zero and $M_i(c)$ tends to infinity.

However this is impossible to realize when $M_i(c)$, $V_i(c)$ have to be replaced by their estimates $\hat{M}_i(c)$, $\hat{V}_i(c)$. For example, even the sample mean estimate $\bar{M}_i(c)$ of $M_i(c)$ defined in (12.2) tends asymptotically ($t \to \infty$) to have a normal distribution with variance given in (12.2), which increases at a rate $\delta(c)$. Also we have the result from the theorem of Karlin [12] that if the density function of $f(y_i)$ of y_i is log convex on (a,∞), $a \geq 0$, then the truncated variance $V_i(c)$ is increasing as c traverses (a,∞). Hence if the underlying distribution of portfolio return belongs to the log convex class, the truncation has the effect of increasing rather than decreasing the original variance. Once again the sampling distribution of the estimators $\hat{\theta} = (\hat{m}, \hat{V})$ of θ, in the presence of truncation c assumes a critical importance.

We consider next the second type of revision problem arising in an intertemporal decision framework. Bertsekas [13] considered a stochastic control model with an N-period planning horizon, where the investor maximizes $E\{U(y_N)\}$, the expected utility of his terminal wealth y_N, when the wealth y_{t+1} in period t+1 follows the dynamic process:

$$y_{t+1} = d_t y_t + \sum_{i=1}^{n} \{(\tilde{r}_{it} - d_t) x_{it}\}$$

$$t = 0, 1, 2, \ldots, N-1$$

containing one riskless asset with a sure rate of return d_t during period t and n risky assets with random returns $\tilde{r}_{it}$ and allocations x_{it} to be determined. Under a certain class of utility functions $U(y_N)$ (e.g., where the reciprocal of the Arrow-Pratt measure of absolute risk aversion is linear in (y_N)), Bertsekas showed that the property of portfolio separation holds in this dynamic framework. This property says that the optimal ratio of the amounts invested in both the risky assets and the single riskless asset (i.e., money) are fixed and independent of initial wealth.

For our optimal revision problem, we first ignore the riskless asset and specify the dynamic process as

$$y_{t+1} = y_t + \alpha_t (\sum_{i=1}^{n} x_{it} \bar{\beta}_{it}) y_t \qquad (13.1)$$

$$t = 0, 1, 2, \ldots, N-1$$

where $\alpha_t y_{t-1}$ is the amount invested at time $(t-1)$ with expected return $\bar{\beta}_t$ which equals $\sum_{i=1}^{n} x_{it} \bar{\beta}_i$, where $\bar{\beta}_i$ denotes the expected value of net return on security i. The decision variables are x_{it} and α_t, which are to be optimally chosen. Note that the dynamic process above can be solved for realized values of $\beta_t = \sum_{i=1}^{n} x_{it} \beta_{it}$ as follows

$$y_t = y_0 \prod_{k=1}^{t} (1 + \alpha_k \beta_k) \qquad (13.2)$$

If x_{it}, β_{it} and α_t did not change over time, the solution (13.2) can be written more simply as

$$y_t = y_0(1 + \alpha\beta)^t$$

for which the linear approximation

$$y_t = y_0(1 + \alpha\beta t) \qquad (13.3)$$

would hold, since α and β are usually small positive fractions. For simplicity we assume that the following approximation

$$y_t = y_0(1 + \alpha_t \beta_t t) \qquad (13.4)$$

$$\beta_t = \sum_{i=1}^{n} \beta_{it} x_{it}$$

holds for (13.2), where y_0 is known, β_i is random and the non-randomized decision variables are α_t, $x_t = (x_{it})$ for $t=1,2,\ldots,N$. Let b_t' denote the row vector $(\beta_{1t}, \beta_{2t}, \ldots, \beta_{nt})$ distributed with a finite mean vector m_t and variance-covariance matrix V_t. Then the mean-variance portfolio model in an intertemporal framework may be formulated as follows:

$$\text{Min } J_N = \sum_{t=1}^{N} [(1+\delta)^{-t} \alpha_t^2 t^2 x_t' V_t x_t]$$

$$\text{s.t. } 1 + \alpha_t x_t' m_t t = c_t \text{ (fixed)} \qquad (13.5)$$

$$\sum_i x_{it} = x_t' e = 1; \; x_t' = (x_{1t}, x_{2t}, \ldots, x_{nt})$$

δ = positive discount rate assumed to be given

Note that for t=1 we get the static mean variance model given by

(1.2) and (1.3) before, by setting $\delta=0$, $\alpha_1 = 1.0$ and $c_1 - 1 = c$. The intertemporal decision model (13.5) may now be solved in two ways. One is by following a myopic decision rule, where we minimize the objective function one time point ahead only, taking care that at each new starting point we allow all the information available up to that point to be used in the estimates of the mean variance parameters. The second way is to solve the complete decision model (13.5) as of now, computing the optimal solutions $\alpha_1^*, \alpha_2^*, \ldots, \alpha_N^*$, $(x_1^*, x_2^*, \ldots, x_N^*)$ for the precommitted horizon. For this optimal calculation we need future values of m_t, V_t for $t=1,2,\ldots,N$. This has to be estimated on the basis of information available as of now. Let us denote these estimates by $m_{t/0}, V_{t/0}$. On using the Lagrangian function as

$$L = \sum_{t=1}^{N} [(1+\delta)^{-t} \, t^2 \alpha_t^2 x_t' V_{t/0} x_t$$
$$+ 2h_t(c_t - 1 - \alpha_t x_t' m_{t/0} t)$$
$$+ 2g_t(1 - x_t'e)]$$

where $2h_t$, $2g_t$ are Lagrange multipliers and e is a vector with each element unity, and applying the necessary and sufficient conditions we note that the optimal vectors x_t^* and scalars $\alpha_t^*, h_t^*,$ and g_t^* must satisfy the following relations for all $t=1,2,\ldots,N$:

$$(1+\delta)^{-1} x_t^* = \frac{(V_{t/0}^{-t} m_{t/0}) h_t^*}{t \alpha_t^*} + \frac{(V_{t/0}^{-1} e) g_t^*}{t^2 \alpha_t^{*2}}$$

CHAPTER IV

$$(1 + \delta)^{-1} \alpha_t^* = \mu_{t/0}/(t\sigma_{t/0}^2);$$

where $\mu_{t/0} = m'_{t/0} x_t^*$, $\sigma_{t/0}^2 = x_t^{*'} V_{t/0} x_t^*$

$$h_t^* = (a_t \gamma_t - \beta_t^2)^{-1} [(c_t - 1) \gamma_t - b_t \alpha_t^*] \qquad (13.6)$$

where $a_t = m'_{t/0} V_{t/0}^{-1} m_{t/0} (1 + \delta)^t$

$$b_t = m'_{t/0} V_{t/0}^{-1} e(1 + \delta)^t, \quad \gamma_t = e' V_{t/0} e(1 + \delta)^t$$

$$g_t^* = (a_t \gamma_t - b_t^2)^{-1} [a_t t^2 \alpha_t^{*2} - t\alpha_t^* b_t (c_t - 1)]$$

$$(1 + \delta)^{-t} \sigma_{t/0}^2 = [t^2 \alpha_t^{*2} (a_t \gamma_t - b_t^2)]^{-1}$$

$$[\gamma_t (c_t - 1)^2 - 2 b_t \alpha_t^* (c_t - 1) t$$

$$+ a_t \alpha_t^{*2} t^2]$$

It is clear from the last relation of (13.6) that one obtains the static efficiency frontier equation (1.4) by setting $\delta = 0$, $t = 1$, $c_t - 1 = c$, $a_t = \alpha$, $\alpha_t^* = 1$, $b_t = \beta$, $\gamma_t = \gamma$ and $\sigma_{t/0}^2 = \sigma_*^2$. The comparative static implications of variations in the estimated parameters for the optimal decision variables may now be easily evaluated. First, the t-period ahead forecasts of the mean variance parameters $m_{t/0}$, $V_{t/0}$ are now required to compute the allocation vector x_t^* and the other decision variables α_t^*. As t gets larger these t-period forecasts tend to become unreliable.

This acts as a deterrent to precommitting oneself to a fixed horizon. Second, the discount rate δ tends to inflate the values of the parameters a_t, β_t and γ_t, which have to be estimated with information available as of time zero. This makes the optimal decision variables less reliable, i.e., they tend to have large standard errors. Note however that the optimal variance $\sigma^2_{t/0}$ of any future t conditional on information available as of time zero is a strictly convex function of $(c_t - 1)$ and hence $\sigma^2_{t/0}$ can be further minimized by choosing a value c^0_t of c_t say. This minimum value is given by

$$\min \sigma^2_{t/0} = (1 + \delta)^{-1}/(e'V_{t/0}e)$$

For large δ this minimal variance tends to zero as $t \to \infty$, provided the estimate of $V_{t/0}$ is still finite. Lastly, one may point out that the estimates of $m_{t/0}$ and $V_{t/0}$ may be obtained in a recursive fashion by followng the Kalman-Bucy filtering technique of control theory [7].

Note however that the approximate formulation (13.5) above does not have any recursive character of modern control theory due to its special structure. Hence we may consider the dynamic process of Bertsekas in a modified version as:

$$y_{t+1} = dy_t + \tilde{b}'x_t + \varepsilon_{t+1} \qquad (14.1)$$
$$y_0 \text{ given}$$

where the row vector $\tilde{b}' = (\tilde{v}_1, \tilde{b}_2, \ldots, \tilde{b}_n)$ is random with finite means and variances and ε_{t+1} is a stochastic disturbance term assumed to be mutually independently distributed with zero mean and constant variance for all t. For simplicity it is assumed that d and b are not time-varying and the allocation vector x_t with elements $x_{1t}, x_{2t}, \ldots, x_{nt}$ is nonrandom. Some plausible forms of objective function used in the literature are as follows.

$$\min_{x_t} E(y_{t+1} - y^0)^2 \qquad (14.2)$$

$$\min_{x_t} J = E(y_{t+1}) - \frac{r}{2} \text{Var}(y_{t+1}) \qquad (14.3)$$

$$\min_{x_t} J = \frac{1}{2} E[\sum_{t=0}^{N-1} \{(y_t - y^0)^2 + s(1 - x_t' x_t)\}] \qquad (14.4)$$

In the first case, i.e., (14.2) and (14.1) we obtain the optimal decision rule (DR):

$$(V_t + b_t b_t') x_t = (y^0 - dy_t) b_t \qquad (15.1)$$

where $b_t = E(\tilde{b}/y_t)$, $V_t = \text{var}(\tilde{b}/y_t)$ are the conditional expectation and conditional variance-covariance matrix of vector $\tilde{b}$, given the observations y_t. This optimal DR may be compared with the deterministic rule known as certainty equivalence rule:

$$b_t b_t' x_t = (y^0 - dy_t) b_t \qquad (15.2)$$

which is obtained by replacing all random variables of the problem by its expected value which may or may not change with new observations. Note that the certainty equivalence rule (15.2) is not cautious, since the variance term V_t is absent. Hence this rule leads to larger variance of control and also of the objective function. Also since the term $b_t b_t'$ is of rank one, it does not define x_t uniquely.

In case of the objective function (14.3) the optimal DR at any future stage t is given by

$$x_t = (rV_t)^{-1} b_t \qquad (15.3)$$

where r is a positive measure of risk aversion in the Arrow-Pratt sense. Like (15.1) it has elements of caution built into the DR: the larger the variance term V_t the smaller the control. Also the higher the risk aversion coefficient r, the lower the optimal control.

Next we consider the intertemporal objective function (14.4) where s is a positive scalar that can be interpreted as a Lagrange multiplier associated with the constraint $\sum_{i=1}^{n} x_{it}^2 = 1$ for all t. By redefining the Lagrangian function

$$L = J + E\{\{\lambda_{t+1}(dy_t + \bar{b}'x_t + \varepsilon_{t+1} - y_{t+1})\}$$

and setting the derivatives $\partial L/\partial t_t$ and $\partial L/\partial x_t$ equal to zero, one obtains

$$E\{y_t - y^0 + d\lambda_{t+1} - \lambda_t\} = \bar{y}_t - y^0 - d\bar{\lambda}_{t+1} - \bar{\lambda}_t = 0 \qquad (15.4)$$

and

$$E\{sx_t + \lambda_{t+1}\tilde{b}\} = 0 \qquad (15.5)$$

The second equation (15.5) involves the product of two random variables λ_{t+1} and $\tilde{b}$. If we assume these to be approximately independent, then we obtain

$$x_t \doteq - \bar{\lambda}_{t+1} b/s \qquad (15.5)$$

Taking expectation of both sides of the process equation (14.1) and on eliminating x_t by using (15.5) we obtain the following set of optimal trajectory equations for $t=0,1,2,\ldots,N-1$:

$$\bar{\lambda}_{t+1} = d^{-1}(\bar{\lambda}_t - \bar{y}_t + y^0)$$
$$\bar{y}_{t+1} = \frac{b'b}{sd}\bar{\lambda}_t + (d - \frac{b'b}{sd})\bar{y}_t + \frac{b'b}{sd}y^0 \qquad (15.6)$$
$$\lambda_N = 0, \ y_0 \text{ and } y^0 \text{ given}$$

These equations are to be solved recursively. A convenient method is to use the so-called Riccati transformation

$$\bar{\lambda}_t = P_t \bar{y}_t$$

and eliminate λ from the equations (15.6). This finally leads to a closed-loop optimal feedback rule

$$x_t = G_{t+1}\bar{y}_t + h_t \qquad (15.7)$$

where G_{t+1} and h_t are suitable time-varying functions, of which

G_{t+1} depends on p_{t+1} used in the Riccati transformation. Dynamic programming and conjugate gradient are some of the computation methods which can be used to solve the optimal feedback rule (15.7) Note that this requires updating G_{t+1} and h_t each time new information becomes available.

Two remarks are in order. First, the feedback rule (15.7) is only approximate, since it used the approximation (15.5). Otherwise it would be nonlinear. Second, one may explore a steady state approximation of (15.7) when G_{t+1} and h_t take a constant value independent of time.

6. Concluding Remarks

The microeconomic problems of how to select an optimally diversified portfolio for a single investor have been very inadequately treated in the current portfolio theory. Moreover when there exist so many mutual funds in the actual market, with each fund like a diversified portfolio there exist for the investor the nontrivial problems of how to choose between different funds and how to revise it over time. The information and advice tendered by the investment brokerage agencies provide some channels by which the investor can be helped in arriving at an optimal decision. How useful has been such advice by the investment brokers? These and related issues need more theoretical and empirical investigations.

CHAPTER IV
References

1. Szego, G.P. Portfolio Theory. New York: Academic Press, 1980
2. Bawa, V.S., Brown, S.J. and R.W. Klein. Estimation Risk and Optimal Portfolio Choice. Amsterdam: North Holland, 1979
3. Beedles, W.L. On the assymmetry of market returns. Journal of Financial and Quantitative Analysis 14 (1979)
4. Black, F. and M. Scholes. The valuation of option contracts and a test of market efficiency. Journal of Finance 27 (1972), 399-417
5. Lehmann, E.L. Theory of Point Estimation. New York: John Wiley, 1983
6. Jacob, N.L. A limited diversification portfolio selection model for the small investor. Journal of Finance 29 (1974), 847-856
7. Sengupta, J.K. Information and Efficiency in Economic Decision. Boston: Martinus Nijhoff Publishers, 1985
8. Sengupta, J.K. A theory of portfolio revision: robustness and truncation problems. International Journal of Systems Science 15 (1984), 805-824
9. Neyman, J. and E.L. Scott. Outlier proneness of phenomena and of related distributions, in J.S. Rustogi, ed. Optimizing Methods in Statistics. New York: Academic Press, 1971
10. Tobin, J. Estimation of relationships for limited dependent variables. Econometrica 26 (1958), 24-36

11. Amemiya, T. Regression analysis when the dependent variable is truncated normal. Econometrica 41 (1973), 997-1016
12. Karlin, S. Some results on optimal partitioning of variance and monotonicity with truncation level, in G. Kallianpur, P.R. Krishnaiah and J.K. Ghosh eds. Statistics and Probability. Amsterdam: North Holland, 1982
13. Bertsekas, D.P. Dynamic Programming and Stochastic Control. New York: Academic Press, 1976

CHAPTER V

Portfolio Efficiency under Singularity and Orthogonality

1. Introduction

Portfolio efficiency has been evaluated in current theory in three different ways. First, we have the efficiency frontier formulation in the Markowitz-Tobin sense, which minimizes the variance of portfolio return subject to a lower bound on expected return. This yields a sequence of vectors of optimal investment proportions, given the mean variance parameters or their statistical estimates. The role of singularity in this framework arises in two ways, e.g., through the near-singularity of the variance-covariance matrix as the size of the portfolio is increased, and through the presence of large standard errors of the estimates of the mean return vector. Szego [1] has recently considered the first case, i.e., the near-singularity of the variance-covariance matrix of returns and its impact on enlarging the size of a given portfolio. Secondly, we may refer to several econometric tests based on the vector of optimal investment proportions, which are designed to evaluate a portfolio for *ex ante* mean variance efficiency. Thus Roll [2] has discussed the following test: Let x_p be the fixed vector of investment proportions for portfolio p and $\hat{x}_p$ be the sample efficient portfolio with mean $\hat{\mu}_p = \hat{m}'x_p$, where prime denotes transpose and $\hat{m}$ is the mean return vector for a sample of N time-series observations on n security returns. Then the statistic $z = \hat{x}_p'\hat{x}_p$ is asymptotically normal with mean $x_p'x_p$ and variance whose sample analog is

$$\hat{\sigma}_z^2 = x_p'(\hat{x}_p\hat{x}_p')x_p/N$$

provided x_p is *ex ante* efficient in the mean variance sense. A critical region is easily constructed in terms of

$$\sqrt{N}(\hat{x}_p'\hat{x}_p - x_p'x_p)/(x_p'\hat{x}_p\hat{x}_p'x_p)^{\frac{1}{2}}$$

which for large samples tends to be distributed like a normal variate N(0,1) with zero mean and unit variance. One basic difficulty in applying this test is that we have to invert a suitable size variance-covariance matrix. If the inverse does not exist, any comparison of alternative portfolio allocations becomes useless. Recently, Sengupta [3] has developed a set of econometric tests for assessing the impact of ill-conditioning and non-singularity of the variance-covariance matrix. An empirical application of these tests to be portfolios constructed from standard stocks traded in the New York Stock Exchange (NYSE) shows very clearly that even in limited diversification portfolios of small sizes the singularity problem may be very important. The third approach to portfolio efficiency evaluation analyzes the performance profile of different mutual funds, where each fund can be viewed as a portfolio composed of securities suitably selected for risk diversification purposes. Thus, Jensen [4] analyzed the time series of returns for 115 mutual funds traded in the New York Stock Exchange over a twenty year period 1945-64 and found little empirical evidence for any individual fund to do significantly better than that expected from mere random chance. Later studies [5-6], although more refined and technical failed to contradict Jensen's finding that the funds are unable to out perform the benchmarks given by the capital asset pricing model (CAPM). However the Jensen critique fails to distinguish between the so-called "balanced" funds and "growth" funds and the fact that the funds themselves are highly positively correlated. The latter implies that a cautious rule of portfolio choice may be adopted by the investor combining different mutual funds into a composite portfolio. Thus, let V_n and V_m be the variance-covariance matrix of returns of two composite portfolio with n and m mutual funds respectively (m<n). Let K be the total number of selections of m funds out of n, where $V_m(k)$ denotes the covariance matrix for any selection k = 1, 2, .., K. For any selection k, the investor considers the maximum risk $\lambda_1(k)$ = $x'(k)V_m(k)x(k)$ in terms of the greatest eigenvalue of $V_m(k)$ and then minimizes $\lambda_1(k)$ over k = 1, 2, ..., K to obtain the minimax risk level λ_1^* and its associated eigenvector $x_{(1)}^*$ say. If each $V_m(k)$ has positive elements and the property of irreducibility, then the allocation vector $x_{(1)}^*$ is unique and has positive elements, which can be so normalized that they sum to unity. The practical relevance of these minimax portfolios denoted here by $x_{(1)}^*$ asumes added importance, when the returns of different mutual funds are highly positively correlated.

We discuss here mainly the static and some dynamic aspects of the singularity problem as it affects optimal portfolio choice. Three aspects of the singularity problem are analyzed in some detail e.g., (1) the non-existence of the inverse of the variance-covariance matrix of returns, (2) the distance measure $D(x(1), x(2))$ of any pair of distinct portfolio allocations of a given size as affected by large standard errors of mean variance parameters, and (3) the failure of normality as a distribution model for the security returns and the effects of skewness and asymmetry on optimal portfolio choice.

2. Mean Variance Model under Singularity

In its simplest verson the portfolio model in the Markowitz-Tobin tradition considers investor who has to optimally invest a proportion of his wealth, x_i in a risky asset i, where $x_i \geq 0$ for $i=1,2,\ldots,n$ and $\sum_{i=1}^{n} x_i = 1$ and the return $\tilde{m}_i$ on asset i is random. Any choice of vector $x' = (x_1, x_2, \ldots, x_n)$ that is feasible i.e.

$$x \geq 0 , \; e'x = 1 \qquad (1.1)$$

where e is an n-element column vector with each element unity and prime denotes transpose, is called a portfolio choice or policy with random return $\tilde{z}$:

$$\tilde{z} = \tilde{m}'x , \; \tilde{m}' = (\tilde{m}_1, \tilde{m}_2, \ldots, \tilde{m}_n) \qquad (1.2)$$

If the return vector $\tilde{m}$ is distributed with mean vector $m = (m_i)$ and variance-covariance matrix $V = (v_{ij})$, then for any fixed non-stochastic decision vector x, the portfolio return $\tilde{z}$ has the mean $E(\tilde{z}|x) = m'x$ and variance $\sigma^2 = \sigma^2(\tilde{z}|x) = x'Vx$.

Given the parameter set (m,V), or their suitable sample estimates, the standard portfolio theory solves for the optimal solution from the following quadratic programming (QP) problem:

$$\min_{x \in R} \sigma^2 = x'Vx \qquad (1.3)$$

where $R: \{x | e'x = 1, m'x \geq \mu ; x \geq 0\}$ (1.4)

and μ is the minimal lower bound on expected returns for any feasible portfolio, which has to be usually preassigned by the investor. By varying the minimal return parameter μ in the decision space, the whole set of efficient portfolios (σ^2, μ), which is also called the efficiency frontier, is determined according to the traditional one-period portfolio theory and its recent extensions.

For our purpose, the model given in (1.3), (1.4) may be considered in a more specialized form. It is clear that the nonnegativity restriction $x \geq 0$ in (1.4) may be dropped, if one is allowed to go short in one security or, take a long position in another. Also, the inequality $m'x$ may be taken as an equality by appropriately preassigning μ or, redefining the set R. As a matter of fact, the portfolio model due to Szego [1], hereafter called model A redefines the set R as follows:

<u>model A</u> $\min_{x \in R} \sigma^2 = x'Vx$, $R = \{x | m'x = \mu, e'x = 1\}$ (1.5)

One must note some important features of this specification. First, the level of μ must be preassigned by the investor.

PORTFOLIO EFFICIENCY UNDER SINGULARITY AND ORTHOGONALITY

Hence if the parameter vector m is empirically estimated, such an equality $m'x = \mu$ may be difficult to preassign exactly. On the other hand, if μ is considered random, one may have to reformulate the equation for portfolio returns. One possible reformulation, closely related to linear regression theory, is as follows:

$$\mu = \tilde{m}'x + \varepsilon = m'x + u'x + \varepsilon \qquad (1.6)$$

where

$$\tilde{m} = m+u \;, \quad \tilde{m} \sim N(m,V)$$

and $\quad \varepsilon \sim N(0,\sigma^2)$

Here the errors u and ε are assumed for simplicity to be normally distributed i.e., $u \sim N(0,V)$, $\varepsilon \sim N(0,\sigma_\varepsilon^2)$ independently of the decision vector x. Two interpretations of the random returns model (1.6) are possible. One is the conditional view, when the decision vector x is nonrandom, i.e. it defines a pure strategy in game theory language. The second allows x to be a mixed strategy, i.e. randomized decision rules. Taking the conditional view of (1.6), one may write the conditional mean and conditional variance of portfolio returns as follows:

$$E(\mu|x) = m'x \;, \quad Var(\mu|x) = x'Vx + \sigma_\varepsilon^2 + 2Cx \qquad (1.7)$$

where $C = E(\varepsilon u')$ denotes the covariance matrix of ε and u'. If ε is statistically independent of u for any x, then the covariance $E(\varepsilon u')$ is zero and hence $Var(\mu|x) = x'Vx + \sigma_\varepsilon^2$. Following the statistical theory of linear regression, we may adopt the following definitions:

Definition 1

A decision vector x_0 is said to be μ_0-unbiased, if $E(\mu|x_0) = \mu_0$ and it has minimum variance (MV) if $Var(\mu|x_0) \leq var(\mu|x)$ for all other feasible vectors $x \neq x_0$. Here feasibility specifies some normalization restrictions like $e'x \leq 1$ or, $x \geq 0$ or, $x'x \geq 1$ as the context makes it clear.

Definition 2

If a feasible decision vector x_0 is not μ_0-unbiased, while the target level μ_0 is preassigned by the investor, then the mean squared error $MSE(x_0)$ associated with vector x_0 is: $MSE(x_0) = Var(\mu|x_0) + (m'x_0-\mu_0)^2$, where the term $(m'x_0-\mu_0)$ indicates the bias.

Definition 3

For any two distinct and feasible decision vectors x_1, x_2 which are not μ_0-unbiased, x_1 is said to strictly dominate x_2 in efficiency, if

$$e_{12} = MSE(x_1) - MSE(x_2) < 0 \qquad (1.8)$$

PORTFOLIO EFFICIENCY UNDER SINGULARITY AND ORTHOGONALITY   197

For weak domination it must hold that $e_{12} \leq 0$.

It is clear that for any biased decision vector, one could decompose the divergence of the portfolio return μ defined in (1.6) from the target value μ_0 into two parts:

$$\mu - \mu_0 = (\mu - E\mu) + (E\mu - \mu_0)$$

where $E\mu = m'x$ and the term $(E\mu - \mu_0)$ specifies the bias component when $E\mu \neq \mu_0$. Expressing the distance (d) as a weighted average

$$d = w(\mu - E\mu) + (1-w)(E\mu - \mu_0), \quad 0 \leq w \leq 1 \qquad (1.9)$$

and taking expectation Ed^2 of its squared value, one obtains

$$Ed^2 = w^2 x'Vx + (1-w)^2 (m'x - \mu_0)^2 \qquad (1.9)$$

this reduces to the MSE criterion if $w = 0.5$ and Ed^2 is multiplied by the scale factor of four. On the basis of this MSE criterion (1.9) one could formulate a portfolio model alternative to (1.5) as follows:

<u>model B</u> $\min f(x) = w^2 x'Vx + (1-w)^2 (m'x - \mu_0)^2 \qquad (2.1)$

where $R: \{x | k(x'x - 1) \geq 0\} \qquad (2.2)$

k: a small positive scalar

w: a nonnegative weight, $0 \leq w \leq 1$

Here it is implicitly assumed that $|m'x-\mu_0| > 0$ so that any feasible vector $x \varepsilon R$ would not be admissible, if it holds $m'x = \mu_0$. Thus one may explicitly include this in the constraint set R as:

$$R: \{x | k(x'x-1) \geq 0; \ m'x \geq \mu_0 \pm c, \ c > 0\} \quad (2.3)$$

Note that the normalization constraint takes the form $k(x'x-1) \geq 0$ instead of the form $1-e'x \geq 0$ used in (1.1), for any set of finite and nonnegative x_i's with at least one positive element can be made to satisfy either of the normalization conditions. A nonnegative scalar k is used for the reason that the sensitivity of a solution will be analyzed for $k \to 0$.

A second feature of the formulation (1.5) in model A is that the optimal solution vector denoted by x_* and the associated objective function $\sigma_*^2 = \min \sigma^2$ may be explicitly computed as a function of μ as follows:

$$\sigma_*^2 = h(\mu) = (\alpha\gamma-\beta)^{-1}(\gamma\mu^2-2\beta\mu+\alpha) > 0$$

$$\quad (2.4)$$

$$x_* = t(\mu) = (\alpha\gamma-\beta)^{-1}[V^{-1}(\gamma m-\beta e)\mu + V^{-1}(\alpha e-\beta m)]$$

where $\alpha = m'V^{-1}m$, $\beta = m'V^{-1}e$, $\gamma = e'V^{-1}e$ and α,β,γ are normally positive and $\alpha\gamma > \beta$, since V is assumed to be strictly positive definite. It is clear that the function $\sigma_*^2 = h(\mu)$ is strictly convex in μ and hence if any specific μ has to be preassigned by the investor, an obvious choice is to minimize σ_*^2 by setting $\mu = \mu_0$.

$$\mu_0 = \beta/\gamma = \mu_* \quad \text{say} \tag{2.5}$$

leading to $\sigma_{**}^2 = \min \sigma_*^2 = 1/\gamma$. For any other choice of μ not equal to β/γ, the variance σ_*^2 will be higher than $1/\gamma$. This analysis needs modification however, if V is not of full rank and/or the parameter m is not completely known. In this case, one may follow the MSE criterion to set up an analogous model:

<u>model C</u> $\min_{x \in R} f = w^2 x'Vx + (1-w)^2(m'x-\mu_0)^2$

where

$$R: \{x \mid m'x = \mu_0 + c, \; x'x \geq 1\}$$

and c is a scalar indicating the degree of bias. For any non-zero c, the optimal solutions $\bar{x}$ of this model (2.6) cannot be μ_0- unbiased. The inequality $x'x \geq 1$ is used only when the covariance matrix V is singular and it has the same interpretation as in model B; otherwise the inequality is assumed to be non-binding (i.e. the associated Lagrange multiplier is zero).

Another feature of model A which seems unrealistic concerns the homogeneity assumption behind the choice set for n securities, each of which is treated on the same basis as it were. In real life, the set of securities is usually divided into say T groups or risk-classes, so that we have a lower bound on expected return for each group of choice. For this case, assume for simplicity that there are n securities in each risk-class or group where there are T risk-classes. Let x be a column vector with nT elements and M a matrix with

T rows and nT columns and μ_0 and c be column vectors each with nT elements. Then a generalized version of model C would be:

model D $\quad\quad \min_{x \varepsilon R} f = x'\tilde{V}x + (Mx-\mu_0)'(Mx-\mu_0)$

where

$$R:\{x|Mx = \mu_0 + c\} \quad\quad (2.7)$$

where the weight coefficient w is dropped and the dimensions of $\tilde{V}$ and e are appropriately enlarged. It is clear that if any element of c is nonzero, then the equality $Mx = \mu_0$ cannot hold and hence the assumption of a nonzero c_i, i=1,2,...,nT implies that the optimal solutions, if any must be μ_0-biased. Thus, for any two feasible decision vectors x_1, $x_2 \varepsilon R$ and a preassigned level of μ_0, one may say that x_1 is <u>strictly</u> more efficient than x_2, if $e_{12} = MSE(x_1) - MSE(x_2) = f(x_1) - f(x_2) < 0$ and x_1 is <u>weakly</u> more efficient than x_2, if $e_{12} \leq 0$ and $x_1 \neq x_2$.

Now consider model B and assume that the set R is non-empty. A case which is most operational takes the normalization constraint as an equality i.e. $k(x'x-1) = 0$ and using λ as the Lagrange multipler the Lagrangean function may be written as

$$L = w^2 x'Vx + (1-w)^2(E\mu-\mu_0)^2 - \lambda k(x'x-1) \quad\quad (3.1)$$

where m'x is denoted by $E\mu$. Let $\bar{x}$ denote an optimal solution with respect to the Lagrangean function, then it must satisfy the first order conditions:

$$[w^2V+(1-w)^2mm'-\lambda kI]\ \bar{x} = (1-w)^2\mu_0 m \qquad (3.2)$$

$$\sigma^2(\bar{x}) = \bar{x}'V\bar{x} = \frac{(1-w)^2}{w^2}\mu_0\bar{\mu} - \frac{(1-w)^2}{w^2}\bar{\mu}^2 + \frac{\lambda k}{w^2} \qquad (3.3)$$

$$\lambda k(\bar{x}'\bar{x}-1) = 0 \qquad (3.4)$$

where $\bar{\mu}$ denotes $m'\bar{x}$ and w ($0 \leq w \leq 1$) is a scalar weight coefficient. Note that if we drop the coefficient w i.e. $f = x'Vx + (m'x-\mu_0)^2$, then for any solution $\bar{x}$, $f(\bar{x}) = MSE(\bar{x})$. In this case, the relations (3.2) and (3.3) become:

$$[V + mm' - \lambda kI]\ \bar{x} = \mu_0 m \qquad (3.2)'$$

$$\sigma^2(\bar{x}) = \mu_0\bar{\mu}-\bar{\mu}^2 + \lambda k \qquad (3.3)'$$

where $\bar{\mu} \neq \mu_0$. Even if V or, $(V+mm')$ is singular, there must exist a value of scalar k, such that the matrix $[V+mm'-\lambda kI]$ is nonsingular. Denoting the solution $\bar{x}$ of $(3.2)'$ by $\bar{x}(k)$ one may obtain:

$$\bar{x}(k) = [V+mm'-\lambda kI]^{-1}(\mu_0 m)$$

This shows that as a function of k, the optimal solution vector $\bar{x}(k)$ may be written into the form of a convergent power series:

$$\bar{x}(k) = \bar{x}(0) + ky_1 + k^2 y_2 + \ldots$$

where $\bar{x}(0)$ satisfies $[V+mm']\bar{x}(0) = \mu_0 m$ and $y_1, y_2, \ldots$ are certain vectors which may be obtained. This suggests that $\bar{x}(k)$

may be closely related to ridge regression in the theory of statistical regression under singular normal equations [7]. Thus, one may postulate normalization conditions differently than in (2.2). For instance, let x_0 be a value desired by the decision-maker and he wishes to compute a biased solution vector such that x is close to the desired vector x_0 i.e. $(x-x_0)'(x-x_0) = \delta$, where δ is a small positive number. In this case, (3.2)' would appear as:

$$(V + mm' - \lambda I)\bar{x} = \lambda x_0 + \mu_0 m \qquad (3.4)$$

From this we see that as a function of λ, the biased solution vector $\bar{x}(k)$ may be expressed as a convergent power series in λ:

$$\bar{x}(\lambda) = \bar{x}(0) + \lambda y_1 + \lambda^2 y_2 + \lambda^3 y_2 + \ldots \qquad (3.5)$$

where $\bar{x}(0)$ satisfies $(V+mm')\bar{x}(0) = \mu_0 m$ and y_1, y_2, y_3 are certain vectors to be computed. Note that we have

$$\lim_{\lambda \to 0} \bar{x}(\lambda) = \bar{x}(0)$$

$$\lim_{k \to 0} \bar{x}(k) = \bar{x}(0)$$

Hence the impact of an increase in λ or, in k on the respective optimal solution vectors $\bar{x}(\lambda)$ or, $\bar{x}(k)$ may be directly computed and evaluated. Thus, it is clear that the property of robustness or stability may be invoked for determining a suitable value of λ in the expansion (3.5) for instance. For

PORTFOLIO EFFICIENCY UNDER SINGULARITY AND ORTHOGONALITY

several small values of λ (or k as the case may be), one may compute the optimal solutions of the minimization model B. Then by running parabolas through these points for each component of $\bar{x}(\lambda)$, we can empirically estimate $\bar{x}(\lambda)$, y_1, y_2.

Several other points may be noted for model B with the Lagrangean function given by (3.1). Denote by $L(k)$ the value of L in (3.1) when the optimal solution vector $\bar{x}(k)$ is used in place of x. The optimal value $\bar{w}$ of w would then be given by minimizing $L(k)$ with respect to w i.e.

$$\bar{w} = (\bar{\mu}(k)-\mu_0)^2 \{(\bar{\mu}(k)-\mu_0)^2 + \sigma^2(k)\}^{-1} \qquad (3.5)$$

where $\bar{\mu} = m'\bar{x}(k)$ and $\sigma^2(k) = \bar{x}(k)'V\bar{x}(k)$. Thus, if $\bar{\mu}(k)$ tends to μ_0, then $\bar{w}$ tends to zero, whereas $\bar{w}$ tends to unity, if $\sigma^2(k)$ tends to zero. Secondly, if w is set equal to unity in L defined by (3.1), the first order condition (3.2) defines an eigenvalue problem in $\theta = \lambda k$:

$$|V-\theta I| = 0, \quad (V-\theta I)\bar{x} = 0 \qquad (3.6)$$

If the rank of V is $m < n$, then after suitable ordering if necessary, the first m eigenvalues θ_i may be ordered as $\theta_1 \geq \theta_2 \geq \ldots \geq \theta_m > 0$, the remaining $(n-m)$ eigenvalues being zero. For minimization of L we must take the smallest eigenvalue θ_m and the associated eigenvector $x(\theta_m) = \bar{x}_m$ say. However, in this case one may also characterize minimax policies [8] by considering for each k the maximum eigenvalue $\theta_1 = \theta_1(k)$

say and the associated eigenvector $x(\theta_1(k)) = \bar{x}_1(k)$. Let K be the domain of variations of $k\varepsilon K$. Then, over K we seek $\min_{k\varepsilon K} \theta_1(k)$ and the associated eigenvector for specifying minimax portfolio policies, which have certain robustness properties.

Third, the role of k in case of singularity of the covariance matrix V needs to be emphasized. It is clear from (3.2)' that if V is singular, so is the matrix (V+mm'); hence for k tending to zero, the solutions $\bar{x}$ become very unstable. This instability can be prevented by choosing a large value of k such that the matrix (V+mm'-λkI) becomes positive definite. If there are several such k's, one chooses the smallest of them to break the instability associated with the singularity of V. If however V is not singular and w satisfies the condition $0 < w < 1$, we may still have the MSE ($\bar{x}(k)$) associated with the solution $\bar{x}(k)$ lower than that (i.e. MSE $\bar{x}(0)$ associated with $\bar{x}(0)$ solved from (3.2)' with k=0.

Theorem 1

Let $\bar{x}(k)$ and $\bar{x}(0)$ be two distinct solutions of the equation (3.2)', corresponding to k positive and zero respectively. Then there must exist a value k* of k and a neighborhood N(k*) such that for all $k\varepsilon N(k^*)$ it holds

$$\text{MSE}(\bar{x}(k)) - \text{MSE}(\bar{x}(0)) = e_{k0} \leq 0$$

Proof

By definition we have $e_{0k} = \bar{\mu}^2(k) - \mu_0\bar{\mu}(k) - \lambda k - c$, where c denotes a constant not dependent on k and $\bar{\mu}(k) = m'\bar{x}(k)$. Since $\bar{\mu}(k)$ is a continuous function of k and e_{0k} is strictly convex in $\bar{\mu}(k)$ for every positive k, there is a value k^* for which $\bar{\mu}(k^*)$ solves the equation $e_{0k} = 0$. By continuity the neighborhood $N(k^*)$ cannot be empty and hence the result.

Corollary 1.1

If w is equal to zero, so that the bias term has no importance in the objective function, then $\sigma^2(\bar{x}(k)) > \sigma^2(\bar{x}(0))$ for every positive value of k. Further, any two eigenvectors of the system $(V-\lambda kI)\bar{x}(k) = 0$ are mutually orthogonal.

Corollary 1.2

If $(1-w)$ is equal to zero, so that the variance term has no effect on the objective function, then there must exist a value k^0 of k and a neighborhood $N(k^0)$ such that it must hold for all $k \in N(k^0)$ that $\bar{x}(k) > \bar{x}(0)$.

Corollary 1.3

If w is a proper positive fraction, $0 < w < 1$, so that both the bias and the variance components have their importance in the investor's objective function, then the minimum MSE value of w denoted by $\bar{w}(k)$ is given by

$$\bar{w}(k) = \frac{\beta^2(k)}{\beta^2(k)+\sigma^2(k)}, \text{ for all nonnegative } k$$

where $\beta(k)$ is the bias component i.e. $|\bar{\mu}(k)-\mu_0|$, $\bar{\mu}(k) = m'\bar{x}(k)$ and μ_0 is the desired target value. If the bias (variance) component is zero, then the optimal weight $\bar{w}(k)$ is necessarily zero (one).

Next we consider model C and write the Lagrangean function as

$$L(c) = w^2 x'\tilde{V}x + (1-w)^2 c^2 - 2\lambda(m'x-\mu_0-c)-k \qquad (4.1)$$

where 2λ and k are Lagrange mulipliers and $\tilde{V} = (V + \frac{k}{w^2} I)$, where k is a suitable nonzero scalar if the covariance matrix V is singular. If V is nonsingular then k is set equal to zero, so that $\tilde{V} = V$. Assume now that the restriction set R in model C is nonempty and denote the optimal solution by $\bar{x}(k)$. Then the following result holds:

Theorem 2

For any nonzero c, the optimal solution $\bar{x}(k)$ has the MSE equal to σ^2+c^2 where $\sigma^2 = (\mu_0+c)^2(m'\tilde{V}^{-1}m)^{-1}$ and k is a fixed nonnegative number. If c is negative, i.e. $c = -\gamma$, $\gamma > 0$ and $\gamma = -c$ is optimally chosen to minimize the MSE, then the minimum MSE denoted by $MMSE(\bar{x}(k))$ equals $(1+s_k)^{-1}\mu_0^2$, where $s_k = m'\tilde{V}^{-1}m$ and the optimal γ equals $\bar{\gamma}(k) = (1+s_k)^{-1}\mu_0$. For $k=0$, $MMSE(\bar{x}(0))$ and $\bar{\gamma}(0)$ take the values $(1+s_0)^{-1}\mu_0^2$ and $(1+s_0)^{-1}\mu_0$ respectively, where $s_0 = m'V^+m$, V^+ being a suitable generalized inverse of V satisfying $VV^+V = V$.

Proof

For a singular variance-covariance matrix V, a suitable positive value of k may be fixed so that $\tilde{V}$ is nonsingular. It is clear that such a choice of k is always possible. If V^{-1} exists then k is set equal to zero and $\tilde{V}$ becomes V. By applying the first order conditions of the Kuhn-Tucker theorem to the Lagrangean function, it follows that

$$MSE(\bar{x}(k)) = c^2 + s_k^{-1}(\mu_0+c)^2$$
$$MMSE(\bar{x}(k)) = (1+s_k)^{-1}\mu_0^2$$
$$\bar{y}(k) = (1+s_k)^{-1}\mu_0$$

For k equal to zero, the inverse of V is not defined, if V is singular; hence we replace the inverse by a generalized inverse, V^+ of V. However the value of s_0 is not unique, since the generalized inverse is not unique.

Corollary 2.1

For any fixed c, the optimal value of w is given by

$$\bar{w} = c^2(\sigma^2+c^2)^{-1}, \quad 0 \leq \bar{w} \leq 1$$

But for the optimal value y, where $c = -y$, $y > 0$ the weight equals:

$$\bar{w}(\bar{y}(k)) = (1+s_k)^{-1}$$

where s_k may be interpreted as the positive square root of the multivariate distance from the zero vector measured by

$$D_k^2 = D_k^2(m,o,\tilde{V}) = m'\tilde{V}^{-1}m$$

For all positive k, $D_k^2 < D_0^2$ since $s_k^2 < s_0^2$. Also $\bar{y}(k)/\bar{w}(\bar{y}(k)) = \mu_0$ indicating that the target level of return expresses the ratio of two weights at the optimum.

Corollary 2.2

If $\bar{y}(k)$ is set equal to the target value μ_0, then σ^2 is zero but the MSE($\bar{x}(k)$) = μ_0^2, which is larger than the MMSE ($\bar{x}(k)$) value $\mu_0^2(1+s_k)^{-1}$ for every positive value of k.

A few remarks on the case of singularity of the covariance matrix V when k is zero may be in order, particularly when the Lagrangean function takes the form

$$L = x'Vx + 2\lambda \; (m'x-\mu_0+\gamma) \qquad (4.2)$$

for a fixed γ say. The necessary conditions for solving for the (n+1) unknowns x and λ are:

$$Vx = \lambda m \qquad (4.3)$$
$$m'x = \mu_0-\gamma \qquad (4.4)$$

where x cannot be uniquely solved from (4.3), since V^{-1} does not exist. Besides the generalized inverse V^+ of V, which is of course not unique, two other methods can be applied for solving for x from the n linear equations (4.3). One method uses the number of independent equations in (4.3) and solves for a reduced system of equations. For instance if the rank of V is $m < n$, then we choose any m columns of V which are linearly independent and express the remaining $(n-m)$ columns as linear combinations of them. Partition $V = (V_B \ V_N)$ where V_B is of order n by m and V_N of n by $(n-m)$. Let $x = (x_B \ x_N)$ be corresponding partitions of x where x_B is of order m by one and x_N of $(n-m)$ by one. Then the equation system (4.3) can be written as

$$V_B \ x_B = b - V_N x_N \ , \ b = \lambda m \qquad (4.5)$$

where there exists a matrix L such that $V_N = V_B L$, since the columns of V_N are linearly dependent on V_B. Also, since rank $V = m$, there exists a matrix D with order n by $(n-m)$, rank $(n-m)$, such that

$$VD = \emptyset \ , \ \text{where} \ \emptyset \ \text{is the null matrix} \qquad (4.6)$$

we have to solve (4.5) under the restriction (4.6) for the m unknowns contained in the basis vector x_B. Since rank $V_B = m$, the linear system can be solved in principle. Note however that there may be more than one choice of m linearly independent columns out of n, $n > m$ and also the choice of matrix D is not unique. The restriction $x'x = 1$ in model C helps to restore some uniqueness in these choices, if we write it as

$(x-x_0)'(x-x_0) = 1$, where x_0 is a vector point to which solutions are required to be close.

A second method for analyzing the case of singularity of V is to adjoin a new set of linear equations $Bx = 0$ to $Vx = b$ so that the coefficient matrix $C = \binom{V}{B}$ has rank n . This requires that we choose a matrix with rank $(n-m) > 0$. Thus, one can seek a solution vector $\bar{x}$ of order n by one which minimizes $(b-Vx)'(b-Vx)$ subject to $Bx = 0$. This leads to the constrained least squares solution as:

$$\bar{x} = (V'V + B'B)^{-1} V'b \qquad (4.7)$$

where the difference $(b-V\bar{x})$ may be interpreted as the error vector ε , where

$$\varepsilon'\varepsilon = b'H'Hb , \quad H = I - V(V'V + B'B)^{-1} V'$$

It is clear that if H is not a null matrix, the error sum of squares $(\varepsilon'\varepsilon)$ can be reduced by suitably modifying the singular matrix V . In the Lagrangean function (4.1) for model C , one could use this modification

$$\tilde{V}^{-1} = (V'V + B'B)^{-1} V'$$

provided a matrix B is available.

Theorem 3

For model D the optimal solution $\bar{x}$ for any nonzero vector c has the mean square error

$$MSE(\bar{x}) = (\mu_0+c)' S^{-1}(\mu_0+c) + c'c$$
where $S = M\tilde{V}^{-1}M'$

If $c = -\gamma$, where γ is a nonnegative vector with at least one positive element and γ is optimally chosen to minimize MSE, then

$$MMSE(\bar{x}) = (\mu_0-\bar{\gamma})' S^{-1}(\mu_0-\bar{\gamma}) + \bar{\gamma}'\bar{\gamma}$$

where the optimal value $\bar{\gamma}$ of γ is:

$$\bar{\gamma} = (I+S)^{-1}\mu_0$$

Proof

The method of proof is straightforward. It is based on the first order conditions applied to the Lagrangean function:

$$L(c) = x'\tilde{V}x + c'c - 2\lambda'(Mx-\mu_0-c)$$

Corollary 3.1

If S is symmetric and positive definite, then $MSE(\gamma_0) - MSE(\bar{\gamma}) = \mu_0'A\mu_0$, where $A = I-(I+S)^{-1}$ is a positive definite matrix and the vector γ_0 is equal to μ_0 but not to $\bar{\gamma}$. In this sense $MSE(\bar{\gamma})$ dominates over $MSE(\gamma_0)$.

Corollary 3.2

If there are no constraints on x in model D, then $MSE(\bar{x})$ takes the value $\mu_0' B \mu_0$, where

$$B = M(\tilde{V}+M'M)^{-1}[\tilde{V}+M'M](\tilde{V}+M'M)^{-1}M'$$

The vector portfolio model, i.e. model D has an interesting application when the covariances for any pair of securities belonging to different risk-classes or groups are zero, i.e. the risk-classes are independent. Let x_t be the allocation vector in the t-th risk-class with a scalar return $y_t = r_t' x_t$, where the return vector r_t is random with mean m_t and covariance matrix V and let p_t denote a screening variable, if the particular class or group is selected or not, i.e. p_t is a nonnegative number, $0 \leq p_t \leq 1$. If $p_t = 1.0$, then the t-th group is totally selected. If $0 < p_t < 1$, then it is fractionally selected and for $p_t = 0$, it is not selected. The mean and variance of $p_t y_t$ then becomes $p_t m_t' x_t$ and $\sigma_t^2 = p_t^2 x_t' V_t x_t$ where $t=1,2,\ldots,T$. A simpler version of model D would then appear as follows:

$$\min_{x \varepsilon R} f = \sum_{t=1}^{T} [p_t^2 \sigma_t^2 + (p_t m_t' x_t - \mu_{0t})^2]$$

where $R: \{x_t, p_t | p_t m_t' x_t = \mu_{0t} + c_t, 1 \geq p_t \geq 0 \quad (5.1)$

for all $t=1,2,\ldots,T\}$

Here μ_{0t} are target returns for each risk-class and c_t indicates the degree of bias. One may also add a condition $\sum_{t=1}^{T} p_t = h$, where only h out of nT securities are to be selected.

For example, if $T=2$, $\mu_{01} = \mu_{02} = \mu_0$, $c_1 = c_2 = c_0$ and $p_2 = 1-p_1$, then the Lagrangean function for the problem (5.1) is

$$L = p_1^2 \sigma_1^2 + (1-p_1)^2 \sigma_2^2 + (p_1\mu_1 - \mu_0)^2 + ((1-p_1)\mu_2 - \mu_0)^2$$

$$- 2\lambda_1(p_1\mu_1 - \mu_0 - c_0) - 2\lambda_2((1-p_1)\mu_2 - \mu_0 - c_0) \qquad (5.2)$$

where we have dropped the constraint on p_t and the notation $\mu_t = m_t' x_t$ has been used. It is clear from (5.2) that for any fixed value of p_1, $0 \le p_1 \le 1$ the above model could be solved for optimal vectors $\bar{x}_1$, $\bar{x}_2$ and the multipliers $\bar{\lambda}_1$, $\bar{\lambda}_2$ say. Given such optimal solutions the constants $\bar{\sigma}_t^2 = \bar{x}_t' V_t \bar{x}_t$, $\bar{\mu}_t = m_t' \bar{x}_t$ may be evaluated. The function $L = L(p_1)$ may then be minimized at the second stage by an optimal choice of p_1. For example, if we minimize the objective function of (5.1) at the optimum, the optimal value of p_1 turns out to be:

$$\bar{p}_1 = \frac{\bar{\mu}_2^2 + \bar{\sigma}_2^2}{D} + \frac{\mu_0 |\bar{\mu}_1 - \bar{\mu}_2|}{D}$$

where $$D = \sum_{t=1}^{2} (\bar{\mu}_t^2 + \bar{\sigma}_t^2)$$

It is clear that the optimal value $\bar{p}_1$ can be zero, only if $\bar{\mu}_1 = \bar{\mu}_2$ and $(\bar{\mu}_2^2 + \bar{\sigma}_2^2) = 0$. The larger the mean square error for second group (i.e. $\bar{\mu}_2^2 + \bar{\sigma}_2^2$), the lower is its weightage in selection, i.e. lower $\bar{p}_2$ and the higher the weightage in rejection.

The vector portfolio model has a second interpretation in terms of the minimax principle, if there is prior information [7] on the allocation vector x, before model D is used. If the prior information is constrained to lie in the convex set

$$C: \{x | x'Ax \leq k\} \tag{5.3}$$

where $k \geq 0$ is a given constant term and A is a known matrix, which is assumed to be positive definite. For minimax estimation problems Toutenberg [7] has considered component by component restrictions like

$$a_i \leq x_i \leq b_i$$

for each element x_i, which occur in economic models where the decision maker has subjective prior knowledge. In this case the matrix A in (5.3) becomes diagonal. If the prior information of type (5.3) has to be adjoined to model D, the restriction set R has to include this set C. Note that such prior information will be almost always true, if the positive value of k is chosen large enough. The larger k, the less binding is the constraint in (5.3), so that for $k \to \infty$, the constraint vanishes altogether. Geometrically, the prior information set (5.3) is an ellipsoid centered at the origin. More generally the decision-maker may center it at x_0 say, when the convex set becomes

$$C = \{x | (x-x_0)'A(x-x_0) \leq k, \ k > 0\}$$

Note however that k is the maximum of $x'Ax$ or, $(x-x_0)'A(x-x_0)$ and hence if we minimize k by rewriting the objective function of (2.7) as

$$\min f = x'Vx + (Mx-\mu_0)'(Mx-\mu_0) + k \qquad (5.5)$$

Denoting by $\tilde{A}$ the matrix A/k, the Lagrangean function for model D may be written as

$$L(x) = L(x,\lambda,r|c,k) = x'Vx + c'c - 2\lambda'(Mx-\mu_0-c) + r(1-x'Ax)$$
$$= x'\tilde{V}x + c'c - 2\lambda'(Mx-\mu_0-c) + r \qquad (5.6)$$

where $\tilde{V} = V - rA$ and λ, r are suitable Lagrange multipliers. Assume that a positive r exists for a suitable k satisfying

$$r(1-x'Ax) = 0 \qquad (5.7)$$

The the following result holds.

<u>Theorem 4</u>

For model D with the Lagrangean function (5.6), the minimax solution $\bar{x}$, for any fixed vector c, has the mean square error

$$MSE(\bar{x}) = (\mu_0+c)'\tilde{S}^{-1}(\mu_0+c), \quad \tilde{S} = M[V - \frac{r}{k}A]^{-1}M'$$

The mean square error $MSE(k)$ viewed as a function of $k \geq 0$ has the optimal property for all positive k :

$$e_{0k} = MSE(0) - MSE(k) = (\mu_0+c)'D(\mu_0+c)$$

where D is a positive definite matrix, provided S, $\tilde{S}$ and $(\tilde{S}-S)$ are positive definite.

Proof

The first part follows from the necessary conditions applied to the Lagrangean function (5.6). For the second part,

$$e_{0k} = (\mu_0+c)'(S^{-1}-\tilde{S}^{-1})(\mu_0+c)$$

If S, $\tilde{S}$ and $(\tilde{S}-S)$ are positive definite, then $D = S^{-1}-\tilde{S}^{-1}$ is positive definite. Hence the result.

Corollary 4.1

If the matrices M, V and A are diagonals with positive diagonal elements, then D is necessarily positive definite. Further, e_{0k} is a strictly concave function for all k.

Corollary 4.2

For a positive k, the derivative $\frac{dMSE(k)}{dk}$ is positive definite (positive semidefinite), if the matrix GAG' is so, where $G = \tilde{S}^{-1}M\tilde{V}^{-1}$.

Theorem 5

If for model D with the Lagrangean function (5.6), λ is zero (i.e. the constraint $Mx-\mu_0-c \geq 0$ is not binding), then the minimax solution $\bar{x}$, for any fixed vector c is the eigenvector of the following generalized eigenvalue equation

$$(V - \frac{r}{k} A) x = 0 \qquad (5.8)$$

If V is diagonalized by the nonsingular transformation $z = W'x$, where $W^{-1}VW'^{-1} = I$ then for minimum $MSE(Z)$ one must choose the maximum eigenvalue of the system:

$$(\theta I - B) z = 0 , \quad B = W^{-1}AW'^{-1} , \quad \theta = k/r$$

where the simple eigenvalues θ_i can be ordered as $\theta_1 \geq \theta_2 \geq \ldots \geq \theta_p > 0$, where p is the dimension of square matrix A.

Proof

By the assumed conditions, the relevant part of the Lagrangean function (5.6) is

$$L(x) = x'(V - \frac{r}{k} A) x$$

this leads to the eigenvalue equation

$$(\theta I - B) z = 0 , \quad z = W'x$$

where the eigenvalues θ_i, $1 \leq 1 \leq p$ are all positive, since V and A are positive definite. For minimum MSE, we have to minimize $z'z$; hence the maximum eigenvalue $\theta^* = \max \theta_i$ and the associated eigenvectors z^* or, $x^* = W'^{-1} z^*$ are to be selected.

Corollary 5.1

The decision vectors z_*, $x_* = W'^{-1} z_*$ corresponding to the minimum eigenvalue $\theta_* = \min_i \theta_i$ lead to maximal variance and maximal MSE.

Corollary 5.2

If V is singular in (5.8) and of rank $s \geq 1$ and A is diagonal with positive diagonal elements, then there are exactly s positive eigenvalues, of which the minimum has to be selected for the minimum of MSE.

3. A Distance Criterion Approach

Now we consider the random portfolio return μ defined in (1.6) before and first take the case

$$\mu = m'x + \varepsilon, \quad \varepsilon \sim N(0, \sigma_\varepsilon^2) \qquad (6.1)$$

where the parameter vector m is known and the decision vector x is nonstochastic. This return equation may be an approximation to (1.6) in the sense that the error $(\hat{m}-m)$ is negligible for all x. Now the investor's goal is to attain a target level $\mu = \mu^o$ by choosing the control vector x. Since μ is stochastic we proceed as follows. Let U^o be an appropriately defined region in the Euclidean space E' which covers the point μ^o, i.e.

PORTFOLIO EFFICIENCY UNDER SINGULARITY AND ORTHOGONALITY 219

$\mu^0 \varepsilon U^0$. Then one chooses the n dimensional control vector x for which the probability that the random point $\mu = m'x+\varepsilon$ will lie inside the region U^0 is maximized. From the assumption of normality of the error component ε in (6.1), it is clear that the region U^0 should be taken to be an ellipse of the form

$$h^0 = (\mu-\mu^0)(\sigma_\varepsilon^2)^{-1}(\mu-\mu^0) \leq c^2 \qquad (6.2)$$

centered at μ^0, where c is a suitable scalar number. If σ_ε^2 is known, the quantity $h = (\mu-m'x)(\sigma_\varepsilon^2)^{-1}(\mu-m'x)$ will have a chi-square distribution with one degree of freedom. Hence the region U^0 can be interpreted as a confidence region for μ at the level α, $0 < \alpha < 1$ in the sense that if we choose the control vector $x = \bar{x}$ such that $E\mu = m'\bar{x} = \mu^0$, where E is expectation, then the probability is $\text{Prob}(\mu \varepsilon U^0) = \alpha$. Thus, the optimal control vector $\bar{x}$ can be obtained by maximizing the likelihood function of the event $\mu^0 = m'x+\varepsilon$, which is of the form

$$L(x) = (2\pi)^{-\frac{1}{2}}(\sigma_\varepsilon^2)^{-\frac{1}{2}} \exp[-\tfrac{1}{2}(\mu^0-m'x)(\sigma_\varepsilon^2)^{-1}(\mu^0-m'x)]$$

This leads to the QP problem:

$$\min_{x \varepsilon R} f(x) = -\log L(x) = \frac{1}{2\sigma_\varepsilon^2}(\mu^0-m'x)^2 \qquad (6.2)$$

where $R: \{x | \sum_{i=1}^{n} x_i = 1, \ x_i \geq 0\}$

Disregarding the constraints for simplicity due to reasons mentioned at the outset, the optimal control vector $\bar{x}$ satisfies the equations

$$mm'\bar{x} = m\mu^0 \qquad (6.3)$$

The coefficient matrix mm' here is singular and hence no unique solution of (6.3) can be obtained. If we introduce a constraint in the form of (5.3) with r as Lagrange multiplier, then (6.3) reduces to

$$(r\sigma_\varepsilon^2 A + mm')\bar{x} = m\mu^0 \qquad (6.4)$$

with the unique solution $\bar{x} = (r\sigma_\varepsilon^2 A + mm')^{-1} m\mu^0$. In contrast to (6.3), this solution in (6.4) is unique and includes the effect of the variance term σ_ε^2, i.e. the larger the variance, the lower the control vector.

Next consider the general model defined in (1.6) before:

$$\mu = m'x + u'x + \varepsilon \qquad (6.5)$$

where it is assumed that u and ε are statistically independent with zero mean and finite variances. By assumption μ is normally distributed with mean $m'x$ and variance $(x'Vx + \sigma_\varepsilon^2)$. Taking logarithm of the likelihood function of the event $\mu^0 = m'x + u'x + \varepsilon$ would now lead to the objective function:

$$\max_{x} \log L(x) = -\tfrac{1}{2} D \qquad (6.6)$$

where

$$D = \log(\sigma_\varepsilon^2 + x'Vx) + (\mu^0 - m'x)^2 (\sigma_\varepsilon^2 + x'Vx)^{-1}$$

Denoting the expression $(\sigma_\varepsilon^2 + x'Vx)$ by the positive scalar t and minimizing $f(x) = \log t + (\mu^0 - m'x)^2 t^{-1}$ subject to $t = \sigma_\varepsilon^2 + x'Vx$, when t is varied parametrically, one could obtain the optimal solution by numerical methods, if necessary. Let there be a positive value of $\bar{r}$ satisfying $r(t - \sigma_\varepsilon^2 - x'Vx) = 0$, where r is the Lagrange multiplier, then the optimal control vector may be obtained from

$$(\bar{r}\bar{t}V + mm')\bar{x} = m\mu^0 \qquad (6.5)$$

where $\bar{t}$ is the value associated with $\bar{r}$ such that

$$\bar{t} = \sigma_\varepsilon^2 + \bar{x}'V\bar{x}, \quad \bar{r}(\bar{t} - \sigma_\varepsilon^2 - \bar{x}'V\bar{x}) = 0$$

It is clear that if $\bar{t}$ is set to unity, then the unique solution $\bar{x}$ denoted by $\bar{x}(\bar{r})$ may be computed from (6.5) as:

$$\bar{x}(\bar{r}) = (\bar{r}V + mm')^{-1} m\mu^0, \quad \bar{r} > 0 \qquad (6.6)$$

Since we have

$$\bar{x}(\bar{r})'V\bar{x}(\bar{r}) = \frac{m'\bar{x}(\bar{r})}{\bar{r}}(\mu^0 - m'\bar{x}(\bar{r}))$$

it is clear that for all $r > \bar{r} > 0$,

$$\bar{x}(r)'V\bar{x}(r) \leq \bar{x}(0)'V\bar{x}(0), \quad \text{if} \quad \mu^0 \geq m'\bar{x}(r)$$

$$\frac{\partial \bar{x}(r)}{\partial r} \leq \frac{\partial \bar{x}(0)}{\partial r}$$

Hence using (6.6), the following result may be easily proved.

Theorem 6

For some value of r, $r \geq \bar{r}$ there must exist a nonempty neighborhood $N(r)$ such that the mean square error $MSE(r)$ associated with $\bar{x}(r)$ has the property

$$\frac{\partial}{\partial r}(MSE(r) - MSE(0)) \leq 0 \text{, for } \mu^0 \geq m'\bar{x}(r)$$

where $MSE(r) = \bar{x}(r)'V\bar{x}(r) + (\mu^0 - m'\bar{x}(r))^2$
$MSE(0) = \bar{x}(0)'V\bar{x}(0) + (\mu^0 - m'\bar{x}(0))^2$

Corollary 6.1

The $MSE(r)$ viewed as a function of k is a convex function of r and hence it can be minimized over the neighborhood $N(r)$.

Some operational implications of the distance criterion approach may now be mentioned in brief. First, the method is applicable when μ in (6.1) is a vector rather than a scalar, e.g. in model D defined in (2.7). For instance, if $\varepsilon = Mx - \mu^0$ is a vector with a multivariate normal distribution $N(0,R)$ with a variance-covariance matrix R, then the optimal control equation for $\bar{x}$, analogous to (6.3) would be

$$(M'R^{-1}M)\bar{x} = M'R^{-1}\mu^0$$

The more general case of (6.5), although more nonlinear could in principle be handled in a similar manner.

Second, the distance criterion based on the likelihood, function of the event μ^0, i.e. $\mu^0 = m'x + \varepsilon$ say, can be applied to any distribution other than the normal, provided the maximum of the likelihood function with respect to the control vector x can be computed. For instance, in portfolio theory, it has sometimes been postulated that the disturbance term ε is not always normally distributed, instead log normal approximation holds better in some cases. If this is so, let $\phi = \log \varepsilon = \log z$, where $z = \mu^0 - m'x$ is assumed to be positive in model (6.1) where $\phi \sim N(0, \sigma_\phi^2)$. Maximizing the logarithmic likelihood of the event ϕ leads to the equations

$$-\frac{m \log z}{z} = 0 \qquad (6.7)$$

which could be solved in principle for the optimal control vector $\bar{x}$. For instance, using the approximation $\log z = z - 1 - \tfrac{1}{2}(z-1)^2$ for $z > 0.5$, one obtains for the optimal approximate solution

$$mm'\bar{x} = m(\mu^0 - a), \quad a = 3 \text{ or, } 1.$$

Note that the equations (6.7) may be called maximum likelihood (ML) equations and for any density function, the ML equations may be denoted as:

$$g(\mu^0 - m'x) = 0 \qquad (6.8)$$

where the distance $(\mu^0 - m'x)$ defines the event for which the log likelihood is maximized. Unlike the standard estimation problem, these ML equations (6.8) have to be solved for the control vector $\bar{x}$. If an optimal solution exists which is exact and defines a unique maximum of the likelihood function, then this is a global solution, otherwise we have a local maximum valid in a specified neighborhood. It is clear that robustness aspects of an optimal solution can be analyzed in this framework by examining the size of the local neighborhood. The main objective why one seeks robustness of an optimal control solution is that it may protect us against a small percentage of outliers or, slight departures from the assumption of a specific probability distribution, e.g. normal. Thus, following Huber [9] one may propose for those cases where the underlying probability density of ε is normal in the middle range but with double exponential tails the following robust form of the $g(\tau)$ function defined in (6.8), i.e.

$$g(\tau) = \begin{cases} -a, & \tau < -a \\ \tau, & |\tau| \leq a \\ a, & \tau > a \end{cases} \qquad (6.9)$$

where $\tau = \mu^0 - m'x$. Here the solution will behave near the tails much differently from that in the middle range.

Third, the distance criterion need not preserve symmetry in the sense that the positive $(\mu^0 - m'x > 0)$ and negative $(\mu^0 - m'x < 0)$ deviations are equally weighted, as in the case of normality with the objective function (6.2). Asymmetry can be allowed through departures from normality as in the robustness analysis mentioned in (6.9), or through specific constraints imposed on the state and control space, before solutions are allowed to be admissible. For instance, an investor may be more concerned with loss ($m'x < \mu^0$), when he intends minimizing $(\frac{\mu^0 - m'x}{\sigma_\varepsilon})^2$ which is equivalent to maximizing $-(\mu^- - m'x)^2 / \sigma_\varepsilon^2$. This objective can be easily realized by maximizing the original logarithmic likelihood function $\log L(x)$ defined in (6.2) but with an additional constraint imposed, i.e.

$$\max_{x \in R} L(x) = -(\sigma_\varepsilon^2)^{-1}(\mu^0 - m'x)^2$$

$$R: \{x \mid m'x \geq \mu^0\} \qquad (6.10)$$

Let $\bar{x}$ be the optimal solution and 2λ the optimal value of the Langrange multiplier associated with the constraint such that

$$\lambda = \begin{cases} 0, & \text{for } m'x > u^0 \\ \text{positive}, & \text{for } m'x \leq \mu^0 \end{cases}$$

then the optimal loss function $C(\bar{x}) = -L(\bar{x})$ has the following implication: If $\lambda = 0$, then $C(\bar{x})$ is positive and the optimal solution $\bar{x}$ is obtainable from

$$mm'\bar{x} = m\mu^0$$

But if $\lambda > 0$, then $C(\bar{x})$ is zero and the optimal solution is given by

$$mm'\bar{x} = m(\mu^0 + \lambda\sigma_\varepsilon^2) \; , \; \lambda > 0$$

As a second example, assume that the error component $\varepsilon = \mu^0 - m'x$ has a two-sided exponential distribution with a density:

$$p(\varepsilon) = \exp(-|\varepsilon|) \; , \; \varepsilon = \mu^0 - m'x$$

then the distance criterion approach calls for minimizing the expression $|\varepsilon| = |\mu^0 - m'x|$ by choosing an optimal control vector.

4. Dynamic Aspects

We may now consider some dynamic situations where the singularity problem may arise. Consider an investor who has wealth y_N at the beginning of the N-th period, part of which is reinvested in a riskless asset with a known return r_0 and the other part reinvested in n securities with returns $\tilde{b}_i = r_i - r_0$ for security $i = 1, 2, .., n$. The dynamic process then takes the form

$$y_{N+1} = r_0 y_N + \tilde{b}'u_N + e_{N+1} \qquad (7.1)$$

where the row vector $\tilde{b}' = (\tilde{b}_1, \tilde{b}_2, \ldots, \tilde{b}_n)$ is random, $u_N' = (u_{1N}, u_{2N}, \ldots, u_{nN})$ is the allocation vector to be chosen by the investor and e_{N+1} is the stochastic disturbance term assumed to be independently distributed with zero mean and constant variance σ^2 for all N. For simplicity it is assumed that r_0 and $\tilde{b}$ are not time dependent i.e., stationary.

PORTFOLIO EFFICIENCY UNDER SINGULARITY AND ORTHOGONALITY

The objective of the investor must now be specified to compute explicit soluations. If he chooses a myopic objective of the form

$$\min E[(y_{N+1}-y^0)^2] \qquad (7.2)$$

where y^0 is the target level of return, then he minimizes the expected value of squared deviations from this target by the control vector u_N. The optimal rule then takes the form

$$(V_N + b_N b_N')u_N = (y^0 - r_0 y_N)b_N \qquad (7.3)$$

where $b_N = E(\tilde{b}|y_N)$, $V_N = \text{var}(\tilde{b}|y_N)$ are the conditional expectation and variance of vector $\tilde{b}$ given the observation y_N. This optimal rule may be compared with another rule, called the certainty equivalence (CE) rule, when we replace the dynamic equation (7.1) by its mean level

$$y_{N+1} = r_0 y_N + b_N' u_N$$

and the objective function (7.2) by

$$\min (y_{N+1} - y^0)^2$$

The optimal CE rule now becomes

$$(b_N b_N')u_N = (y^0 - r_0 y_N)b_N \qquad (7.4)$$

Note that the coefficient martix $(b_N b_N')$ in (7.3) is singular, hence the CE rule (7.4) does not determine the optimal allocation decision uniquely. Suppose however that the investor chooses to use an intertemporal objective function of the form

$$J = \tfrac{1}{2} E\left[\sum_{N=0}^{T-1} \{(y_N - y^0)^2\} \right] \qquad (7.5)$$

along with the system dynamics at its mean level:

$$\bar{y}_{N+1} = r_0 \bar{y}_N + \bar{b}_N' u_N \qquad (7.6)$$

where $\bar{y}_N$ is the expected value of y_N, $\bar{b}_N = E\{b | y_N\}$ and $V_N = \mathrm{var}\{b | y_N\}$. The optimal CE rule now becomes

$$[V_{N-1} + \bar{b}_{N-1} \bar{b}_{N-1}'] u_{N-1} = (y^0 - r_0 \bar{y}_{N-1}) \bar{b}_{N-1}$$

Two points are clear. The singularity of the coefficient martix $(\bar{b}_{N-1} \bar{b}_{N-1}')$ is no longer a problem, since a nonsingular variance-covariance matrix V_{N-1} would define the optimal trajectary $\{u_{N-1}, N=1, 2, \ldots, T-1\}$ uniquely. Also, the term V_{N-1} emphasizes very clearly the cautious nature of this optimal allocation policy i.e. the higher the level of V_{N-1}, the lower the level of optimal allocation policy u_{N-1}. Secondly, with increasing information the conditional variance-covariance matrix V_{N-1} has to be updated. This updating process itself may involve singularities of different types that are discussed in various techniques of optimal filtering e.g., Kalman-Bucy filters, Wiener

PORTFOLIO EFFICIENCY UNDER SINGULARITY AND ORTHOGONALITY

filters etc. in the theory of optimal control.

The multiperiod case of th linear-quadratic model (7.5) and (7.6) has the convenient structure that only mean-variance parameters enter into the optimal decision rule i.e. skewness of the returns distribution does not affect the optimality conditions. In a different framework Arditti and Levy [10] have examined the case where the skewness of multiperiod returns of a portfolio directly affects the first three moments of the portfolio's single period return. Denote by $1+y_i$ the i-th one-period return on a given portfolio, where y_i is the rate of return for this period. The first three moments of the distribution of one-period returns are then

$$E[(1+y_i)] = 1+\mu_i$$
$$E[(1+y_i) - (1+\mu_i)]^2 = \sigma_i^2$$
$$E[(1+y_i) - (1+\mu_i)]^3 = \mu_{3i}$$

The N-period return is given by $\prod_{i=1}^{N}(1+y_i)$ and assuming independence over time of the returns, the first three moments of this multiperiod return denoted by 1+M, V and S respectively may be written as

$$1+M = \prod_{i=1}^{N}(1+\mu_i)$$

$$V = \prod_{i=1}^{N}[(1+\mu_i)^2 + \sigma_i^2] - \prod_{i=1}^{N}(1+\mu_i)^2$$

$$S = \prod_{i=1}^{N} [(1+\mu_i)^3 + 3(1+\mu_i)\sigma_i^2 + \mu_{3i}]$$

$$- 3 \prod_{i=1}^{N} (1+\mu_i) \left\{ \prod_{i=1}^{N} [(1+\mu_i)^2 + \sigma_i^2] \right\}$$

$$+ 2 \prod_{i=1}^{N} (1+\mu_i)^3$$

It is clear that the multiperiod skewness measure S would not be zero, even though the one-period skewness μ_{3i} is equal to zero for all values of $i=1,2,\ldots,N$. This explains very clearly the point noted by Arditti and Levy that the monthly data on portfolio returns show little or no skewness, whereas annual portfolio returns are empirically found to be positively skewed. The practical implication of this result is that skewness is a function of the investment horizon which is determined by investor's behavior. If we restrict ourselves to those investors who have short investment horizons, skewness of portfolio returns can be ignored, since existing empirical evidence indicates that. But for investors with longer horizons, the skewness may be a significant factor in the investment decision-making process.

It is clear therefore that the linear-quadratic model (7.5), (7.6) may not prove to be a good approximation when the investment horizon is longer. Other generalized types of dynamic portfolio models involving a chance-constrained formulation of capital and liquidity constraints have been considered by Sengupta [11] in order to evaluate the reliability of the linear decision rule as a basis of approximation. As is expected, the longer the horizon, the less reliable the linear decision rules and this is found mainly due to the significant impact of the skewness parameter.

5. Concluding Remarks

The problem of singularity has both theoretical and empirical dimensions. Theoretically it complicates the problem of computing an efficiency frontier in the mean-variance space, for an individual investor. This is more so if the investment horizon is longer. This may be the reason why limited diversification portfolios have been empirically found to perform better than either the larger portfolios or, random-allocation portfolios. Empirically, the presence of singularity makes it very different to compare alternative portfolio allocations, each having a nearly singular covariance-structure of its returns. This problem needs more investigations both theoretical and empirical.

CHAPTER V

REFERENCES

1. Szego, G.P. Portfolio Theory. New York: Academic Press, 1980

2. Roll, R. Testing a portfolio for ex-ante mean-variance efficiency, in E.J. Elton and M.J. Gruber eds. Portfolio Theory: Twenty Five Years After. Amsterdam: North Holland, 1979

3. Sengupta, J.K. Econometrics of portfolio risk analysis. International Journal of Systems Science, 15(1984), 1023-1037

4. Jensen, M.C. The performance of Mutual Funds in the period 1954-64. Journal of Finance 23(1968), 389-416

5. Copeland, T.E. and J.F. Weston. Financial Theory and Corporate Policy. Reading, Massachusetts: Addison-Wesley Publishing, 1983

6. Berman, O., Modiano, E. and J.A. Schnabel. Sensitivity analusis and robust regression in investment performance evaluation. International Journal of Systems Science 15(1984), 475-480

7. Toutenberg, H. Prior Information in Linear Models. New York: John Wiley, 1982

8. Sengupta, J.K. A minimax policy for optimal portfolio choice. International Journal of Systems Science 13(1982), 39-56

9. Huber, P.J. Robust Statistical Procedures. Philadelphia, Pennsylvania: Society for Industrial and Applied Mathematics, 1977

10. Arditti, F.D. and H. Levy. Portfolio efficiency analysis in three moments: the multiperiod case, in H. Levy and M. Sarnat eds. Financial Decision Making Under Uncertainty. New York: Academic Press, 1977.

11. Sengupta, J.K. Optimal Decisions Under Uncertainty: Methods Models and Management. New York: Springer-Verlag, 1985

CHAPTER VI

Diversification and Robustness in Portfolio Investment: An Empirical Analysis

1. Introduction

Diversification of investment is widely accepted as a useful means of reducing risk of portfolio returns. Recently the diversification issue has attracted some attention due to several theoretical and empirical reasons. From a practical viewpoint the rapid growth of financial assets controlled or managed by institutional investors such as pension funds and investment trusts has generated a demand for realistic standards for judging portfolio performance. The problem of risk measurement has acquired some importance in portfolio evaluation. An individual investor may no longer be satisfied by comparing his individual portfolio returns with a market index, since the market portfolio may have, in all likelihood, very different risk characteristics than the portfolio in question. Existing measures of risk assessment, though useful in several respects have been subject to two basic criticisms. One is that the risk level of a portfolio is assumed to remain inflexible over time, i.e., stationary and the usual mean-variance model of efficient portfolio frontier has excluded the role of skewness of returns distribution altogether. The time-series data on returns of mutual funds have evidenced flexibility of the risk levels of portfolios and this seems quite reasonable since the portfolio managers of mutual funds can be expected to adjust portfolio risk exposure in anticipation of market changes [1,2]. The skewness of returns measured suitably has been found to be statistically significant by a number of researchers [3,4], although it has not been incorporated into the mean-variance model of efficiency frontier, except by an attempt due to Arditti and Levy [5], who showed that the longer the investor's decision horizon, the stronger the impact of skewness on the mean and variance parameters in a multi period case. Since the impact of market information on individual investors is not perfect, it is quite reasonable to expect that the information market is not uniformly efficient either across securities or, over time. The asymmetry of market returns needs therefore to be more thoroughly examined.

The empirical data on common stock returns certainly show that diversification pays up to a limit but the larger the investment horizon the smaller is the merit of diversification. Likewise certain portfolios which are either diversified by industry groups or constructed as limited diversification portfolios (LPD) are found to perform better than portfolios which are diversified randomly. It is clear that the diversification objective is not always captured very adequately by a variance-minimizing portfolio. It seems we need more than one criterion to evaluate diversification. This raises the issue of robustness as another criterion. Robustness can of course be viewed in alternative ways, some of which would be discussed later. In practical terms it indicates a qualitative aspect of diversification which is sometimes termed as "volatility", measured for example by the range or spread between the maximum and minimum observed return. Mutual fund portfolios have been classified by Wiesenberger on the basis of the stated objective (e.g., growth of capital, income) and volatility in relation to the overall market (measured for example by a composite index like Standard and Poor's 500 Index) as follows: code 0 for growth funds (high volatility), 1 for income fund, 2 for balanced fund (average volatility) and code 3 for growth plus income funds. An empirical analysis of 341 mutual fund portfolios for the decade 1959-69 performed by Campanella [6] shows that the four risk classes classified by Wiesenberger fit very clearly the average time-series behavior of the funds. In other words, those investment companies classified as having a generally high volatility do in fact deal in securities with high volatilities. Conversely, for those mutual funds (e.g., balanced funds) where assigned volatilities are considered average, portfolio managers seem to weight investment in individual securities so as to achieve a portfolio volatility close to that of market volatility.

For some mutual funds, e.g., growth funds, portfolio volatility is also found to be non stationary. This non stationarity has two important implications for the investor. One is that the measurement of ex-post average volatility may not be sufficient for assessing risk. The second implication is that a longer horizon is critical in managing such funds in anticipation of market swings and hence with greater opportunities for reward or loss.

Thus investors with a long term horizon of 9 to 10 years can expect on average a volatility consistent with the fund's objective and stated risk class. By contrast the concept of risk class by itself as classified by Wiesenberger may be practically meaningless to investors with short or intermediate term horizon, since the horizon time may not be of sufficient length to include at least one full market cycle together with appropriate response in portfolio volatility. Just as the skewness parameter plays a very active role in portfolio efficiency in a long-run horizon, the capital growth objective influences only those investors who have a long-term outlook. It is clear therefore that the principle of random selection of mutual funds without considering their risk classes would be very inefficient. In other words it is easy to construct a composite portfolio made up of mutual funds stratified by the four risk classes, which will perform better than a randomly selected portfolio. The time-series data on mutual funds may thus be profitably used to construct optimal portfolios which may be superior to the random portfolio. Our empirical attempts below are designed to generate such optimal portfolios and provide econometric tests for their superiority.

The following types of empirical analysis are attempted here:

A. Diversification of common stock investments through "limited diversification portfolios" (LDP)

B. Portfolio performance of mutual funds

C. Evaluation of mutual fund portfolios

D. Estimation of portfolio efficiency frontier

This analysis purports to illustrate the different features of diversification and robustness in optimal investment allocations.

2. Diversification in Small Size Portfolios

Diversification through small size portfolios of common stocks is often advocated on two grounds. One is that the portfolios which are diversified by industry groups tend to perform better than those diversified randomly. Second, it is easier to provide econometric tests to see if enlarging the portfolio improves its performance or not. As Roll [7] has pointed out that the standard econometric tests based on the vector of investment proportions are asymptotically valid only in large samples and they require the inverting of a variance-covariance matrix. For large matrices such tests are almost impossible to apply. Furthermore we have the reason that while the diversification principle is useful for a risk averse investor who intends to minimize overall risk, it does rarely provide specific guidelines to the investor who strives to determine the size and composition of his portfolio. This is particularly true for a small investor who has to optimally choose a subset of a larger set of securities. This choice of subsets has been called "limited diversification portfolios" (LDP) by Jacob [8], who showed on the basis of empirical simulations that diversification beyond eight to ten securities may be superfluous. As small investors dealing in LDPs how should we econometrically compare two portfolio efficiency frontiers, when the estimates are used for the mean variance parameters? How should we revise a given size portfolio in a statistically significant way? Partial answers to these questions are attempted here.

A different situation for extending the diversification principle arises when the investor has to consider robustness of his optimal portolio choice. If k out of n securities (k < n) are to be selected, then we have K selections where $K = \binom{n}{k}$ is the combination of n securities taken k at a time. Each selection may minimize the overall variance of portfolio returns and is therefore diversified; yet not all selections are robust. Also, if the variance-covariance matrix $\hat{V}$ is nearly singular, the principle of minimizing the overall variance of portfolio returns does not make any sense.

Here we consider two data sets of securities currently traded in New York Stock Exchange (NYSE) and perform econometric tests on the effects of diversification, robustness and near singularity of small size portfolios. These tests are simple to apply and are likely to be useful for the small investor in his choice of LDPs.

2.1 <u>Models and Data</u>

Two variants of a portfolio model are used, where the objective function is the same but the constraints are different. The common objective is to minimize the overall variance of returns within a portfolio of securities. The first variant, called model A, utilizes the formulation due to Szego [9] as follows:

$$\underset{x}{\text{Min}} \ \sigma^2 = x'Vx$$

subject to (1.1)

$$m'x = c, \ e'x = 1$$

where x is an n-element column vector representing allocations of investment, e is a vector with unit elements and (m,V) are the mean vector and variance-covariance matrix of the vector of returns $\tilde{r}$. The scalar c denotes minimal return, that is subjectively determined by the investor. Some differences of this formulation from the standard portfolio (e.g., Markowitz) model are to be noted. First, there is no non-negativity restriction on the allocation vector as in Markowitz theory; hence this model (1.1) can be solved by straightforward calculus methods, without adopting quadratic programming algorithms. As a matter of fact, the optimal solutions x_*, σ_*^2 of model A can be explicitly written as

$$x_* = (\alpha\gamma - \beta^2)^{-1} V^{-1} \{(m\gamma - e\beta)c + (e\alpha - m\beta)\} \quad (1.2)$$

$$\sigma_*^2 = \sigma^2(x_*) = \min \sigma^2 = (\alpha\gamma - \beta^2)^{-1}(\gamma c^2 - 2\beta c + \alpha) \quad (1.3)$$

where $\alpha = m'V^{-1}m$, $\beta = m'V^{-1}e$ and $\gamma = e'V^{-1}e$. By varying c on the positive axis one can generate the whole portfolio efficiency frontier for a fixed parameter set m,V.

Second, more than one stage of optimization can be easily built into it. Thus, suppose a small investor has to select k out of n securities (k < n) for reasons of economizing on transactions costs. Then we have K total selections, where

K is given by the combination of n things taken k at a time. Using index s = 1,2,...,K to denote any fixed selection, the optimal solutions of the LDP model

$$\text{Min}_{x(s)} \sigma^2(s) = x'(s) V(s) x(s)$$

subject to (1.4)

$$m(s)' x(s) = c, \ e'x(s) = 1$$

can be easily written out on the analogy of (1.2) and (1.3) as follows:

$$\sigma_*^2(x) = (\alpha(s)\gamma(s) - \beta^2(s))^{-1} \{\gamma(s)c^2 - 2\beta(s)c + \alpha(s)\} \quad (1.5)$$

Now one can minimize $\sigma_*^2(s)$ over the set $s \in K$ of possible selections. Thus, for every fixed s we minimize the overall variance of return for portfolio of size k and then we choose an optimal s over the set K of total possible selections.

Due to the existence of two stages of optimization, other choices like robust selection procedures are also conceivable. For example, a minimax portfolio policy discussed by Sengupta [10] considers first the worst possible risk level (i.e. $\max_x \sigma^2(s)$ instead of $\min \sigma^2(s)$) for each fixed s and then chooses the best of the worst from the set K of total selections. This minimax portfolio policy ($\min_s \max_x \sigma^2$) is robust in the sense that if the estimates $\hat{m}$, $\hat{V}$ of the parameters m, V contain large estimation errors, the bounds of which are not precisely known, then it affords some protection against too large a variance or risk.

The second variant of portfolio policy, called model B has the same objective function as (1.1) but specifies the constraints by the normalization condition:

$$\sum_{j=1}^{n} x_j^2 = x'x = 1.0 \qquad (2.1)$$

which has two interpretations. One is through prior informations [11] when the component by component restrictions on each x_j as

$$a_j \leq x_j \leq b_j, \quad j = 1, 2, \ldots, n$$

can be written as $x'Dx \leq 1$, where D is a suitable diagonal matrix. Geometrically, the prior information set is an ellipsoid centered at the origin, although any other central or desired value may be chosen as a reference. The second interpretation is that it is a normalization condition just like $x'e = 1.0$.

A minimax policy according to model B is then given by:

$$\min_{s} \max_{x(s)} \sigma^2 = x'(s) V(s) x(s)$$
$$\text{subject to} \quad x'(s)x(s) = 1.0 \qquad (2.2)$$

At the first stage we compute the worst possible risk level i.e. $\max \sigma^2 = \sigma^2(x(s))$ and then select in the second stage that $s \in K$ which minimizes the worst possible risk. This may thus characterize a robust policy when estimation errors are present in the sample statistics for the parameters m(s) and V(s).

The minimax policy for model B has two differences from that for model A. First, the maximum eigenvalue $\lambda^*(s)$ of V(s), for each fixed s specifies the worst possible level of risk in model B and it is much easier to compute. Taking the minimum over $s \varepsilon K$ i.e. $\min_{s \varepsilon K} \lambda^*(s) = \lambda_0$ say, specifies the minimax risk and associated with this eigenvalue there is an eigenvector, say x_0, which gives the optimal allocation vector. The portfolio return associated with x_0 may be computed as $m'x_0$. If $m'x_0$ happens to equal or exceed the preassigned level c, then the first constraint of model A will also be satisfied; otherwise we would have a different value of c associated with $m'x_0$. A second feature of minimax policy under model B arises when we consider only positively correlated stocks. In this case the variance-covariance matrix V has positive elements and if it satisfies a condition of indecomposability, then by Perron-Frobenius theorem, every eigenvector $x^*(s)$ associated with the maximum eigenvalue $\lambda^*(s)$ has positive elements and hence the elements of $x^*(s)$ may be so normalized as to sum to unity. Thus the minimax eigenvector x_0 can also be suitably redefined so that its positive elements add to unity. Note also that instead of <u>minimax</u> policy specified by (2.2), we could characterize a <u>minimin</u> policy as:

$$\min_{s} \min_{x(s)} \sigma^2 = x'(s) \, V(s) \, x(s)$$
$$\text{subject to} \quad x'(x) \, x(x) = 1.0 \tag{2.3}$$

where the minimum over selections $s \in K$ of the minimal eigenvalues $\lambda_*(s)$ would characterize the minimin risk level. However, Perron-Frobenius theorem would not be applicable to the minimum eigenvalue $\lambda_*(s)$ and the associated eigenvector, even if $V(s)$ contained only positively correlated stocks.

The specification of minimin policy in (2.3) according to model B may be compared with that in (1.5) for model A. It is clear from (1.5) that the minimal variance $\sigma_*^2(s)$ is a strictly convex function of c, since $\gamma(s)$ is strictly positive by the positive definiteness of each $V(s)$. Hence one may obtain the value c_* of c which minimizes the minimum risk level $\sigma_*^2(s)$ further:

$$c_* = \beta(s)/\gamma(s), \text{ assuming } \beta(s) > 0$$
$$\min_c \sigma_*^2(s) = 1/\gamma(s) \qquad (1.6)$$
$$\min_{s \in K} \{\min_c \sigma_*^2(s)\} = \min_s \{1/\gamma(s)\}$$

We may now describe the framework of the two data sets over which the two optimal portfolio models are empirically applied. The data set I was so constructed as to have a well diversified portfolio with fifteen securities selected from the following industries: the electronics, building materials, petroleum, food products, manufacturing and other service sectors of the U.S. economy e.g., Polaroid, Bank of America, Exxon, Johnson and Johnson, Merrill Lynch, Boeing,

CHAPTER VI

Boise Cascade, Hewlett Packard, Heublin, Marriott, Goodyear, General Electric, General Motors, General Food and Getty Oil. Three of 15 securities i.e. Heublin, GM and Goodyear were dropped because of negative mean returns. Once the 12 securities were selected, quarterly stock prices and dividend data for each stock were recorded for the years 1976-80 excluding the two years 1981 and 1982, which had witnessed recessionary shifts in the economy. NYSE stock reports and Moody's Dividend Records were used to construct quarterly returns r_{it} for stock i and quarter t defined as follows:

$$r_{it} = [(p_{i,t+1} - p_{i,t}) + d_{it}]/p_{i,t+1} \qquad (3.1)$$

where $p_{i,t}$ = price of stock i in quarter t and d_{it} = dividend. In cases where stock splits occurred, the size of the dividends for the following quarters were increased (i.e. adjusted) by the proportion of the split.

The 12 securities are ranked from highest to lowest according to their mean returns (Table 1) and the following portfolios are constructed: portfolio I with all 12 securities, and II with top 8 securities.

TABLE 1.
Mean Returns and Standard Deviations for Data Set I

		Mean	Std. Dev.			Mean	Std. Dev.
1.	Boeing	0.148	0.190	8.	Merrill Lynch	.033	.156
2.	Exxon	.097	.060	9.	General Electric	.030	.045
3.	Getty Oil	.071	.121	10.	Bank of America	.030	.081
4.	Marriott	.067	.105	11.	General Foods	.016	.087
5.	Boise Cascade	.056	.087	12.	Polaroid	0.004	0.189
6.	Johnson & Johnson	0.051	0.107				
7.	Hewlett Packard	0.042	0.116				

The second portfolio of 8 top securities is enlarged by one security chosen from the remaining four. This constitutes four portfolios IIA, IIB, IIC and IID respectively, each of size nine. Two portfolio scenarios are examined. The first compared three different sized portfolios where the number of observations per security is held constant at 19. The second scenario is designed to test a method for selecting an additional security for inclusion in portfolio II with 8 top securities and 19 observations.

The second data set considered eleven securities from Moody's Handbook over the years 1970-81 and a returns series was constructed on the basis of definition (3.1). The mean returns and standard deviations are as follows:

TABLE 1.
Mean Returns and Standard Deviations
for Data Set II

		Mean	Std. Dev.			Mean	Std. Dev.
1.	Ex-cell-o Corporation	0.092	0.254	7.	Holly Sugar	0.176	0.572
2.	Cyclops	.093	.228	8.	Koppers Co.	.204	.287
3.	Greyhound	.059	.219	9.	McLean Trucking	.054	.368
4.	Johnson & Johnson	.116	.266	10.	Norris Industries	.253	.560
5.	Dayton Power & Light	.032	.166	11.	Perkin-Elmer Corporation	0.208	0.495
6.	Dana Corporation	0.177	0.324				

Five classes of portfolios are constructed, each having different degrees of diversification. Portfolio I considers only positively correlated stocks of sizes 4 to 7; there are four such portfolios IA through ID, one of each size from 4 to 7. Group II contains portfolios IIA through IID, each of size 7, where the variance-covariance matrix of returns is suspected to be nearly singular. Portfolio group III includes six small size portfolios, each of size 5 selected randomly from the original 11 securities - i.e. the selections of IIIA through IIIF. In portfolio group IV we have two intersecting portfolios IVA, IVB where the point of intersection plays a critical role in the stochastic dominance of one portfolio over the other. Finally, portfolio group V shows (i.e. VA through VE) the effect of enlarging the base portfolio of 5 securities with serial numbers 1,2,3,6 and 10 by adding the securities bearing serial numbers 4,5,7 and 8.

Data set II has some basic differences from data set I

as follows: First, the securities in data set II are not ranked by their mean returns and eight of the 11 securities were selected randomly from another study related to security market line. This data set II has more diversity in terms of several characteristics e.g., singularity, positively correlated returns, intersecting portfolios and the effects of diversification through increased sizes. Second, while data set I was utilized for model A, the second data set was used for both model A and model B. Thus the effects of constraints like normalization and the lower bound on average portfolio returns may be more readily compared and contrasted for data set II.

The estimation of parameters for both data sets has ssumed normality of the returns series and unbiased maximum likelihood estimates are used for the calculations. To the extent that the normality assumption holds only asymptotically, our econometric tests are asymptotic in the large sample sense.

2.2. Econometric Tests and Implications

The statistical tests applied here are intended to detect for different optimal portfolios, of varying sizes if their optimal risk levels differ in a statistically significant sense. A reference portfolio is selected in terms of which other optimal portfolios are statistically compared; a suitable null hypothesis (H_0) is set up along with the alternative hypothesis (H_1) and tests are performed to see if the null hypothesis can be rejected at the chosen level of significance,

$\alpha = 0.05$, say. If the alternative optimal portfolios do not significantly differ from the reference portfolio in levels of optimal risk, then one has to look for other criteria like robustness to make a best selection. Such criteria are also useful in optimally enlarging a small size portfolio by adding new securities.

2.3 Analysis of Data Set I

Three sets of statistical tests are applied to analyze the differences between optimal portfolios under data set I. First, one notes that the ratio of optimal variances for any pair of portfolios within groups I, II and III, adjusted for degrees of freedom is distributed asymptotically as F distribution and hence the similarity between the three portfolio groups may be statistically tested by comparing the observed value of F with the critical value of 2.40 at 0.05 level of significance. At the three levels of minimal return $c = 0.10$, 0.15, 0.25 whch are most realistic, these test results reported in some detail in Table 3A are as follows:

	I and II	II and III	III and I
F(0.25)	2.209	6.647*	14.681*
F(0.15)	1.406	4.201*	5.906*
F(0.10)	1.188	1.579	1.875

* denotes F to be significant at 0.05 level

DIVERSIFICATION AND ROBUSTNESS IN PORTFOLIO INVESTMENT 249

Thus at the low level of c = 10% none of the three portfolio groups are statistically significant at 5% level. However at levels of c = 0.15 and 0.25, portfolio I and II are significantly different from the smallest size portfolio group III. Portfolios I and II are not statistically different at the three levels of c. Thus in terms of reduction of variance risks, increasing the size of the portfolio beyond eight (i.e. group II) has very little statistical relevance. This tends to provide statistical support to the empirical hypothesis put forward by Jacob referred to earlier. However one should note that these F tests are predicated on a particular level of c preassigned. If the decision maker is not preassigning any particular level of minimal return c, then one should compare for different portfolios the optimal risk return curves. One statistic which is convenient for this purpose is the integrated variance (I_j) defined as

$$I_j = \int_{c_1}^{c_2} \sigma_*^2(I) dc \qquad (3.2)$$

which provides a measure of the area under the curve $\sigma_*^2 = h(c)$. This statistic is closely related to similar measures proposed in time series models where the spectral densities are not directly comparable due to their instability at different points. The results of the integrated variance tests, reported in Table 3B for the most realistic levels of c ($0.137 \leq c \leq 0.310$) show that the portfolio groups II and III are statistically different from I at the 5% level of significance; also II and III are not significantly different at all. Thus, on an overall basis portfolio I stands significantly different from the others. Similar tests applied to the four portfolios in group II, as reported in Table 4 show that there is no statistical difference of the portfolios IIA, IIB, IIC from II.

There is a third characteristic represented by the reciprocal of γ, which is the minimum of the optimal risk levels σ_*^2, in terms of which the optimal portfolios groups can be compared. Denoting by $v = 1/\gamma$ this minimal variance statistic, it follows by the normality assumption that its standard error is $(2/N)^{\frac{1}{2}}v$ where N is the number of observations in the sample. Hence for any two independent portfolios with risk levels v_i and v_j, one could define the t-statistic:

$$t = (v_i - v_j)/[\tfrac{2}{N}(v_i^2 + v_j^2)]^{\frac{1}{2}} \qquad (3.3)$$

If the portfolios are not statistically independent, then this t ratio holds only approximately. The results of these t-tests are as follows:

Portfolios

	I	II	III	IV	IIB	IIC	IID
v	.0008	.0018	.0029	.0010	.0016	.0012	.0194
Std. error	.0003	.0006	.0009	.0003	.0005	.0004	.0063

t(I & II) = 1.492, t(II & III) = 1.049, t(III & I) = 2.210

As in the case of integrated variance statistic, portfolio I turns out to be significantly different from portfolio III at 5% level of the t-test. However at 1% level none of the three portfolios I, II and III are statistically different. This shows, somewhat contrarily to Jacob's finding mentioned earlier, that in terms of overall measures like the I and v-statistic the different optimal portfolio frontiers are rather insensitive to moderate changes in size. To some extent this may be due to the normality assumption made here.

The comparative characteristics of five optimal portfolios reported in some detail in Table 4 may be utilized to test the hypothesis of optimal revision i.e. if by adding one security the existing portfolio can be improved. It is clear that by the integrated variance statistic, only portfolio IID is significantly different from II, while IIA, IIB and IIC are not. However if we compute the F statistics at three specific levels of minimal return $c = .25, .15, .10$ then none of the four revised portfolios IIA-IID is significantly different from II. This shows very clearly that any comparison of optimal portfolio risks by conventional statistical criteria is unlikely to be very useful. One needs suitable other criteria like robustness and adaptivity to new information. Some idea of robustness may be obtained from this data set, if we refrain from making the assumption of normality, or of any distribution for that matter. This becomes then a situation of decision making under complete uncertainty and criteria like minimax may be invoked. For each portfolio within the realistic range of c, $0.137 \leq c \leq 0.310$ one first determines the maximum level of optimal risks and then chooses a portfolio having the lowest level of maximum risks. This determines a best-of-the-worst-risk portfolio i.e. minimax portfolio.

Portfolio	Values of σ_*^2 for c at			Max Row
	0.137	0.209	0.310	
IIA	.0033	.0083	.0200	.0200
IIB	.0032	.0089	.0237	.0237
IIC	.0033	.0086	.0217	.0217
IID	.0028	.0070	.0189	.0189
II	.0036	.0120	.0360	.0360
Min column	.0028	.0070	.0189	

It is clear that IID provides a robust portfolio by the minimax criterion. The level $\sigma_*^2 = 0.0189$ for IID also provides a saddle point in pure strategies, when nature is choosing her strategies c = 0.137, 0.209, 0.310 from a distribution unknown to the decision-maker.

2.4 Analysis of Data Set II

The second data set is in a sense more suitable for analysis of robustness and other characteristics like near singularity of the variance-covariance matrix of portfolio returns. This is because we have here a more diverse group of portfolios, some

DIVERSIFICATION AND ROBUSTNESS IN PORTFOLIO INVESTMENT 253

with all positively correlated stocks (group I) and others with linearly dependent securities (group II). Also both models A and B are applied here for optimal portfolio choice. It is therefore possible in this framework to statistically test specific hypotheses about portfolios belonging to a particular group. Some of these specific hypotheses are as follows:

H_1: The minimax portfolio ID in group I is not significantly different from the others in the same group in its level of optimal risk. If true, this implies that ID has no particular advantage in terms of robustness.

H_2: The quasi-singular portfolios in group II are not statistically stable in terms of variances and not statistically different among one another. If true, this implies that the search for minimal variance portfolio within this class is unproductive, if not meaningless.

H_3: The small size LDPs in group III do not differ in optimal risk levels in a statistically significant sense. If true, this implies a strong case for random selection of portfolios within this group, as has been frequently suggested in current literature.

CHAPTER VI

H_4: The two intersecting portfolios in group IV do not significantly differ in optimal risk, although one dominates the other in terms of risk to the right of the point of intersection.

H_5: Portfolio revision by increasing the size in group V does not lead to the selection of a lower risk portfolio compared to a representative portfolio such as VC, which denotes the minimax portfolio. If true, this makes the case for random selection stronger.

For portfolios in group I the minimax policy given by (2.2) in model B can be directly evaluated from the eigenvalue equations:

$$(\hat{V}(s) - \lambda_s I) x(s) = 0 \qquad (4.1)$$

where $\hat{\lambda}_s^*$ and $\hat{\lambda}_{s*}$ are the maximum and minimum eigenvalues of the estimated variance-covariance matrix $\hat{V}(s)$, at which we have the maximum and minimum levels of risk respectively. Since the elements of $\hat{V}(s)$ are maximum likelihood estimates, it follows that the ratio $[\log \hat{\lambda}_s - \log \lambda_s]/(2/N)^{\frac{1}{2}}$ tends to be asymptotically distributed like a unit normal variate for a large N, provided the population parameters λ_s are nonvanishing. On using this theorem the difference in eigenvalues can be statistically tested by the student's t-ratio as follows:

$$\frac{\log \hat{\lambda}_s(1) - \log \hat{\lambda}_s(2)}{(2/N)^{\frac{1}{2}}} \sim t \qquad (4.2)$$

When tested over the eigenvalues for group I portfolios reported in Table 5 the difference in maximum eigenvalues between the minimax portfolio ID and the three others IA through IC are not statistically significant by the t-test at 5% level of significance. Thus, within this group the portfolio ID has no specific advantage in terms of robustness.

The near-singularity of portfolios in group II can be tested in its two aspects: one is to test if the minimum eigenvalue is close to zero in a statistical sense and the other to test the degree of ill-conditioning d defined by the positive square root of the ratio of maximum and minimum eigenvalues. If d is larger than one, we have nonorthogonality of data points. The larger the index d, the worse the ill-conditioning. The actual values of d were

	Portfolios			
	IIA	IIB	IIC	IID
Value of d:	27.67	24.33	28.05	21.01

It is clear that the degree of ill-conditioning in the sense of departure from orthogonality is very high for the four portfolios in group II. However the ill-conditioning by itself may be neither necessary nor sufficient to produce collinearity of the data points. Hence the need for applying the first test. But for the first test we cannot apply the t-test defined before in (4.2) since the parameter λ_s is zero here. We therefore adopt a different procedure due to Tintner [12] as follows: Let p be the order of the estimated variance-covariance matrix $\hat{V}$, where the eigenvalues of $\hat{V}$ are ordered as

$$\hat{\lambda}_1 \geq \hat{\lambda}_2 \geq \ldots \geq \hat{\lambda}_p$$

Then $\chi^2 = \hat{\lambda}_{j+1} + \cdots + \hat{\lambda}_p$ is asymptotically distributed like chi-square with $(p - j)(N - j - 1)$ degrees of freedom, where N is the total number of observations. With $j = p - 1$ this criterion can be applied to the minimum eigenvalue for each of the portfolios in group II. In each case the null hypothesis that the minimal eigenvalue is equal to zero cannot be rejected at the 5% level of the chi-square test. For example, the portfolio IIA has $\hat{\lambda}_p = 0.0007$, which is much lower than the value of 11.07 for the chi-square statistic at .05 level and 5 degrees of freedom.

The estimated parameters of the six small size portfolios reported in Table 7 suggest a pattern of similarity and closeness in the shapes of the risk return curves $\sigma_*^2 = h(c)$ denoting the efficiency frontier. The minimal levels of optimal risk given by the reciprocal of $\gamma(s)$ for low values of return say $c = 0.10$ the minimal level of optimal risks $\sigma_*(s)$ is lowest

(0.021) for s = IVA and highest (0.027) for IIIB, the difference being statistically insignificant. When we apply however the eigenvalue test (4.1) based on model B, we observe a striking contrast. Portfolio IIIA, which is a minimax portfolio is significantly different from all other portfolios in this group by the one-tailed t-test at 5% level with the exception of IIIC. The detailed results are as follows:

	IIIB	IIIC	IIID	IIIE	IIIF
Computed t value	1.813*	1.742	1.814*	1.813*	2.692*

$t_{.05,11} = 1.796$; * denotes significance

Thus it is clear that the hypothesis H_3 cannot generally be accepted. Some other properties of the minimax portfolio may also be noted. The stocks included in portfolio IIIA are all positively correlated, so by the Perron-Frobenius theorem the nonnegativity of the investment allocation vector is assured. Second, in terms of the levels of minimal variance $\sigma_*^2(s)$, the minimax portfolio IIIA has the lowest value for c = 0.10, .14, 0.16 and 0.20 and next to lowest for c = 0.12 and 0.18.

The two intersecting portfolios IVA and IVB under the hypothesis H_4 define the intersection point c_0 at which the two optimal risk-return curves are equal i.e.

$$\sigma_*^2(IVA) = 0.843c^2 - 0.016c + 0.110$$
$$= \sigma_*^2(IVB) = 0.770 - 0.007c + 0.012$$

In terms of the t-test for the maximum eigenvalues, the two portfolios are not statistically different at 5% level, although IVA dominates in risk over IVB to the right of the intersection point c_0. Also in terms of the reciprocal of $\gamma(s)$ reported in Table 8 the two portfolios are statistically indistinguishable.

For optimally revising a portfolio through enlarging its size, the portfolios in group V offer perhaps the most interesting case of comparison. Taking the minimax portfolio VC as the reference (since it may be robust under incomplete information or uncertainty), one may ask if other portfolios in this group differ significantly from the reference. There certainly do exist other portfolios having optimal risk levels lower than that of the reference - e.g., in terms of the reciprocal of $\gamma(s)$ reported in Table 9, VD and VE have lower optimal risks. Also if we compare different portfolios in this group by the integrated variance statistic I_j defined in (3.2) there exist other portfolios better than VC and this difference is statistically significant at the 5% level of F test. Thus there exists in this case a statistically significant way of revising the reference portfolio VC so as to lower the optimal risk levels in the sense defined.

In conclusion we may say that the statistical tests for comparing alternative optimal portfolios belonging to a specified group have illustrated very clearly how easy it is to overrate the benefits of diversification or to mistakenly believe the difference in risks of two quasi-singular portfolios to be large, when it is vitiated by large standard errors. It is clear that due to the presence of large estimation errors the difference between portfolios considered to be optimal by some criterion may not be statistically significant. If anything this calls for a more detailed investigation of the usual mean variance analysis of portfolio theory as applied to decision problems of the small investor. The assumption of normality also need be tested, since for nonnormal situations statistical comparison of portfolio returns and risk levels become much more difficult.

TABLE 3A:
Comparison of Optimal Portfolio Risks
with Varying Sizes
(Data Set I)

Characteristics	Portfolio Group I (size 12)	Portfolio Group II (size 8)	Portfolio Group III (size 4)
α	7.08	5.68	3.37
β	41.80	48.86	33.15
γ	1241.65	564.57	344.52
Value of c (min. return)	$\sigma_*^2(I)$	$\sigma_*^2(II)$	$\sigma_*^2(III)$
0.25	0.0091	0.0201	0.1336
.15	.0032	.0045	.0189
.10	.0016	.0019	.0030
.05	.0009	.0027	.0147
.025	.0008	.0044	.0309
0.025	0.0009	0.0055	0.0417

Statistical test at | Hypotheses tested

c = 0.25

$H_0: \sigma_*^2(II) < \sigma_*^2(I)$ $H_0: \sigma_*^2(III) < \sigma_*^2(I)$

$H_1: \sigma_*^2(II) \geq \sigma_*^2(I)$ $H_1: \sigma_*^2(III) \geq \sigma_*^2(I)$

$F(0.25) = \dfrac{0.0201}{0.0091} = 2.209;$ $F(.25) = 14.681$

c = 0.15

$F(0.15) = \dfrac{.0045}{.0032} = 1.406;$ $F(.15) = 5.906$

c = .10

$F(0.10) = .0019/0.0016 = 1.188;$ $F(.10) = 1.875$

critical F at 0.05 level = 2.40

TABLE 3B:
Statistical Comparison of Three Optimal Portfolio Groups
(Data Set I)

	Portfolio Group I (size 12)	Portfolio Group II (size 8)	Portfolio Group III (size 4)
Statistics for Comparison	$I_1 = \int_{}^{0.310} \sigma_*^2(I)dc$ $= 0.137$ $= 0.0013$	$I_2 = \int_{}^{0.310} \sigma_*^2(II)dc$ $= 0.137$ $= 0.0058$	$I_3 = \int_{}^{0.310} \sigma_*^2(III)dc$ $= 0.137$ $= 0.0093$

	Test 1	Test 2	Test 3
Tests of alternative hypotheses	$H_0: I_2 < I_1$ $H_1: I_2 \geqq I_1$ $\dfrac{0.0058}{0.0013} = 4.427*;$	$H_0: I_3 < I_1$ $H_1: I_3 \geqq I_1$ $\dfrac{0.0093}{0.0013} = 7.099*;$	$H_0: I_3 < I_2$ $H_1: I_3 \geqq I_2$ $\dfrac{0.0093}{0.0058} = 1.603$

Critical value of F at $\alpha = 0.05$ level is 2.40

TABLE 4:
Comparison of Five Optimal Portfolios in Group II
(Data Set I)

Portfolio Variances σ_*^2

Characteristics	II (size 8)	IIA (size 9)	IIB (size 9)	IIC (size 9)	IID (size 9)
α	5.685	5.723	5.877	5.726	6.905
β	48.857	44.826	45.564	45.719	47.002
γ	564.574	994.579	621.101	804.821	567.393
Value of c (min. return)					
0.25	.0201	.0120	.0139	.0132	.0110
.15	.0045	.0040	.0039	.0040	.0033
.10	.0019	.0018	.0019	.0018	.0019
.05	.0027	.0010	.0018	.0013	.0021
.025	.0044	.0017	.0025	.0016	.0029
0.0125	.0055	–	–	–	–
Integrated Variance (I)	0.0058	0.0024	0.0034	0.0028	0.0016
Tests of Alternative Hypotheses		$H_0: I_{2A} < I_2$	$H_0: I_{2B} < I_2$	$H_0: I_{2C} < I_2$	$H_0: I_{2D} < I_2$
		$\frac{.0058}{.0024} = 2.4167$;	$\frac{.0058}{.0034} = 1.7059$;	$\frac{.0058}{.0028} = 2.0714$;	$\frac{.0058}{.0016} = 3.625$

Critical F at α = 0.05 level is 2.40

DIVERSIFICATION AND ROBUSTNESS IN PORTFOLIO INVESTMENT 263

TABLE 5:
Parameters of Portfolio Group I (Data Set II)

Portfolio	Eigenvalue max	Eigenvalue min	Normalized Eigenvector for max. eigenvalue						α	β	γ
IA	.536	.001	.03	.06	.09	.05	.17	.34	.90	-1.03	85.5
IB	.530	.001	.03	.06	.10	.18	.36	.26	.89	-.67	78.7
IC	.509	.004	.03	.07	.21	.40	.29		.54	3.49	28.6
ID	.508†	.014	.06	.21	.42	.30			.40	2.67	23.9

For minimax eigenvalue †, $\sigma_*^2 = 10.05c^2 - 2.24c + 0.16$

TABLE 6:
Parameters of quasi-singular portfolios (Data Set II)

Portfolio	Eigenvalue max	Eigenvalue min	α	β	γ	σ_*^2
IIA	.536	.0007	.905	-1.027	85.5	$1.12c^2 + 0.03c + 0.01$
IIB	.533	.0009	1.251	.961	86.0	$0.81c^2 - 0.02c + 0.01$
IIIC	.472	.0006	1.194	.855	89.8	$0.84c^2 - 0.02c + 0.01$
IIID	.309	.0007	1.296	.414	86.7	$0.77c^2 - 0.01c + 0.01$

Note: Near singularity is indicated by the minimum eigenvalue not statistically different from zero.

TABLE 7:
Comparison of Six Small Size Portfolios (Data Set II)

Portfolio	max	min	$\alpha(s)$	$\beta(s)$	$\gamma(s)$	$\sigma_*^2(s)$
IIIA	.144†	.007	1.067	.730	74.4	$0.94c^2 - .02c + .01$
IIIB	.423	.004	.645	.267	76.9	$1.51c^2 - .01c + .01$
IIIC	.344	.010	.919	1.020	77.5	$1.10c^2 - .03c + .01$
IIID	.419	.008	.839	3.292	51.6	$1.59c^2 - .20c + .03$
IIIE	.413	.004	.991	2.009	60.3	$1.08c^2 - .07c + .02$
IIIF	.427	.004	.884	.950	77.7	$1.15c^2 - .03c + .01$

† minimax eigenvalue

TABLE 8:
Parameters of Intersecting Portfolios in Group IV
(Data Set II)

	Portfolio IVA				Portfolio IVB		
c	σ^2_*	Max Eigen-value (λ^*)	Normalized Eigenvector for λ^* (column)	c	σ^2_*	Max Eigen-value (λ^*)	Normalized Eigenvector for λ^* (column)
0.10	0.018	0.472	.036	.10	0.018	0.309	0.025
.12	.021		.040	.12	.022		.083
.14	.025		.066	.14	.026		.112
.16	.030		.119	.16	.030		.140
.18	.036		.049	.18	.035		.046
.20	0.042		.382	.20	0.041		.171
			.308				0.423

$\sigma^2_* = 0.843c^2 - .016c + .011$

$\alpha = 1.194, \beta = 0.855, \gamma = 89.8$

$\sigma^2_* = 0.770c^2 - .007c + .0012$

$\alpha = 1.296, \beta = 0.415, \gamma = 86.7$

TABLE 9:
Effects of Enlarging Portfolios in Group V
(Data Set II)

Portfolio	Maximum Eigenvalue	σ_*^2 equals	α	β	γ
IVA	0.418	$2.98c^2 - .25c + .02$	0.437	2.41	57.54
VB	0.433	$1.97c^2 - .34c + .02$	1.693	13.82	161.08
VC	0.411†	$0.94c^2 - .11c + .01$	1.748	12.11	213.51
VD	0.442	$0.66c^2 - .07c + .01$	2.067	10.94	217.82
VE	0.446	$0.55c^2 - .09c + .01$	26.057	280.24	3240.72

† minimax eigenvalue

3. Portfolio Performance of Mutual Funds

The investment performance of mutual funds has drawn increased attention from financial theorists and investment experts in recent times. Since a mutual fund seeks to provide a portfolio with a balanced mixture of securities in terms of risk and return, its performance may be measured in at least three different ways. One is to measure the ability of a fund manager to increase returns through successful prediction of future security prices. The second is to assess the manager's ability to minimize the amount of insurable risk through efficient diversification of the fund portfolio. The third is to perform a comparative evaluation of different mutual funds with regard to their optimal risk and return along their efficiency frontiers. In his pioneering empirical work Jensen [13] considered the investment performance of 115 mutual funds in the U.S. over a 20 year period 1945-64 and applied the first type of measurement of fund manager's ability to predict. His conclusions were two-fold. First, none of the funds were on average able to outperform a naive policy of buying and holding the market portfolio. Second, no individual fund was able to do significantly better than that which could be expected from mere random chance. Recently, Berman, Modiano and Schnabel [14] applied the modern techniques of sensitivity analysis and least absolute residuals regression to a random sample of ten mutual funds selected from the set of 115 studied by Jensen and their results did not contradict Jensen's earlier conclusions.

Our objective here is to apply the remaining two dimensions of performance evaluation, e.g. efficiency in diversification and robustness of selected portfolios. On the basis of the same ten mutual funds selected earlier by Berman, Modiano and Schnabel, we construct eight portfolios, which are different combinations of the ten funds emphasizing income, growth and other objectives. The efficiency in diversification and robustness of performance are then evaluated for these eight portfolios over the period 1945-64.

3.1 Diversification tests and database

The ten mutual funds analyzed here had their specific objectives listed in codes by Weisenberger Financial Services as follows: 0 for growth, 1 for income, 2 for balanced and 3 for growth and income. Thus growth funds emphasized growth of capital much more than income or balanced funds. These objectives seem to reflect the differing risk attitudes of different classes of investors.

DIVERSIFICATION AND ROBUSTNESS IN PORTFOLIO INVESTMENT

	Name of Fund	Objective Code	Sample Size	Returns Mean	Variance
1.	Composite Fund, Inc.	2	16	0.094	0.010
2.	Istel Fund, Inc.	2	14	.104	.013
3.	Massachusetts Life Fund	2	15	.088	.007
4.	National Securities - Dividend Series	1	10	.118	.034
5.	Television Electronics Fund, Inc.	0	12	.141	.032
6.	Group-Securities- Aerospace Science Fund	0	14	.101	.080
7.	Incorporated Investors	3	19	.089	.038
8.	Mutual Investment Fund, Inc.	2	18	.067	.015
9.	United Science Fund	0	15	.123	.032
10.	The George Putnam Fund of Boston	2	18	0.093	.013
	Average			0.102	0.027

The correlation matrix of returns of these randomly selected funds is given in Table 1, which shows very clearly that the funds are highly positively correlated. The growth funds however differ from the balanced funds in two distinct ways. First, they have much higher variance of returns and a higher spread between the lowest and highest variance. Second, the growth funds are more strongly correlated among themselves than with the balanced funds.

These characteristics suggest that one can test the efficiency in diversification for these funds in two ways. One is to construct a set of portfolios on the basis of the ten mutual funds and derive the risk-return efficiency frontier. The objective here is to detect if the efficiency frontiers for the different portfolios so constructed are significantly different in a statistical sense. The following set of eight portfolios is constructed:

$A = (2,4,5,9)$, $B = (2,4,6,9)$, $C = (1,2,4,6)$, $D = (1,2,6,10)$, $E = (1,6,7,10)$, $F = (1,3,7,10)$, $G = (3,7,8,10)$, and $H = (1,2,3,8,10)$.

Of these, three portfolios B, C and H may be considered as typical in emphasizing growth objectives, income plus growth and balanced mixture of objectives. The efficiency frontiers are then derived on applying the following optimizing principle:

$$\min_{x} \sigma^2 = x'Vx \quad\quad \text{subject to} \quad m'x = c, \; e'x = 1 \quad (1.1)$$

where x is an n-element column vector (n=9) denoting proportional allocations of investment x_i (i=1,2,...,8) in eight portfolios, e a vector with unit elements, prime denotes transpose and the parameters (m,V) are the mean vector and the variance-covariance matrix of the random returns for eight portfolios. The optimization model (1.1) above minimizes the total risk measured by variance subject to a minimal level of expected return $c > 0$. Let x_* be the optimal solution for any fixed positive level of c and the minimum value of the objective function be $v_* = \sigma_*^2$. Then one could easily derive the risk-return efficiency frontier as:

$$v_* = (\alpha\gamma - \beta^2)^{-1} (\gamma c^2 - 2\beta c + \alpha)$$
$$= k_0 - k_1 c + k_2 c^2 \quad (1.2)$$

where $\alpha = m'V^{-1}m$, $\beta = m'V^{-1}e$, $\gamma = e'V^{-1}e$ and $k_0 = \alpha(\alpha\gamma-\beta^2)^{-1}$, $k_1 = 2\beta(\alpha\gamma-\beta^2)^{-1}$, $k_2 = \gamma(\alpha\gamma-\beta^2)^{-1}$.

A second way to test the efficiency of the diversification objective realized by the ten mutual funds is to postulate that the variance of returns is almost identical for the different funds and the investor faces the decision problem of selecting the best of the ten normal distributions of returns. Here it is assumed that we have ten normal distributions of fund returns $N_i(\mu_i, \sigma_i^2)$ with means μ_i and a common variance $\sigma_1^2 = \sigma_2^2 = \ldots = \sigma_{10}^2 = \sigma^2$ and the investor's objective is to identify one of these ten funds as the best one, i.e. which produces the highest return. Here we can order the funds from highest to lowest average returns and then apply the recent technique of optimal selection among K normal populations developed by Gibbons, Olkin and Sobel [15]. According to this selection rule we analyze the entire parameter space of μ-values and separate out those regions where we have a strong preference for making a correct selection from those regions where we are indifferent between two or more different selections. The former region is termed the preference zone and the latter the indifference zone. If we order the μ values as

$$\mu_{(1)} \leq \mu_{2} \leq \cdots \leq \mu_{(k)}$$

and define a distance measure $\delta_j = \mu_{(k)} - \mu_{(j)}$, $j = 1, 2, \ldots, k-1$ then the preference zone is where $\delta_j \geq \delta^*$, where δ^* is a specified positive parameter denoting a threshold value. To determine the best normal population, i.e. the population with the highest average return $\mu_{(k)}$, we order the sample mean returns as

$$\bar{r}_{(1)} \leq \bar{r}_{(2)} \leq \cdots \leq \bar{r}_{(k)}$$

and assert that the sample mean $\bar{r}_{(k)}$ corresponds to the population with the largest value of μ i.e. $\mu_{(k)}$. To estimate the true probability of a correct selection of the best population we estimate the population values δ_j by the sample estimates

$$\hat{\delta}_1 = \bar{r}_{(k)} - \bar{r}_{(1)}, \hat{\delta}_2 = \bar{r}_{(k)} - \bar{r}_{(2)} \cdots , \hat{\delta}_{k-1} = \bar{r}_{(k)} - \bar{r}_{(k-1)}$$

and define for the jth sample $j=1,2,\ldots,k$ the statistic

$$\hat{\tau}_j = \hat{\delta}_j \sqrt{n_0}/s \qquad (2.1)$$

where n_0 is the common sample size and s^2 is an unbiased estimate of the common variance σ^2. We may use these $\hat{\tau}_j$ values to estimate the true probability of a correct selection of the best population, i.e. a lower bound P_L and an upper bound probability P_U. Then (P_L, P_U) provides an interval estimate of the true probability of a correct selection. A point estimate P_E lying between P_L and P_U can also be computed by the cluster method suggested by Gibbons, Olkin and Sobel. The computed estimate of P_E must at least be greater than $0.5 + 0.5/k = 0.55$ for $k=10$ for the decision maker to have confidence in the best population selected.

DIVERSIFICATION AND ROBUSTNESS IN PORTFOLIO INVESTMENT

Thus the selection of the best among ten normal populations ordered pairwise in terms of the $\hat{\delta}$-statistic is guided in final analysis by the estimate P_E of the true probability of a correct selection. Unlike the method of the efficiency frontier specified for portfolios of funds in (1.2), the optimal selection and ranking procedure seeks to identify a mutual fund with a highest average return and to estimate the true probability of a correct selection.

3.2 Comparison of risk-return frontiers

The efficiency frontier equation (1.2) for risk-return trade-off may be empirically analyzed in three different ways. One is to specify the minimum variance (v_*) for each portfolio for different levels of the minimal expected return parameter c. This shows the sensitivity of v_* in respect of c. A second way is to test if for any two portfolios the quadratic functions $v_* = k_0 - k_1 c + k_2 c^2$ intersect for a positive value of c within the realistic range $0.6 \leq c \leq 0.14$. If they intersect say at a value c_0, then one portfolio may dominate the other for a value of $c > c_0$. In other words, these portfolios may be very dissimilar on the right of c_0. Thirdly, if the parameter k_2 is positive, the risk-return frontier functions $v_* = v_*(c)$ will be strictly convex and hence the variance v_* may be further reduced by choosing a value c_* of c at which $\partial v_*/\partial c$ is zero. It is clear from (1.2) that

$$c_* = \beta/\gamma \text{ and } v_{**} = \min v_* = 1/\gamma, \quad \gamma > 0 \qquad (1.3)$$

The different portfolios can be compared in terms of the statistic v_{**}. This comparison is meaningful only if the associated c_* value is within the observed range $0.06 \leq c_* \leq 0.14$.

The various statistics of the risk-return efficiency frontiers are reported in Table 2 at the end, which show a broad similarity in curvature except perhaps two portfolios A and G. For portfolio A, the efficiency frontier is strictly concave in c, rather than convex, thus implying that the minimal variance v_* attains its maximum at $c^* = \beta/\gamma = 0.115$. For portfolio G, the minimal variance v_* reaches its lowest value zero say (although in this case it is negative at -0.0026) at a high value of c_* close to 19.3%.

For the three typical portfolios B (growth), C (income and growth) and H (balanced) the sensitivity of minimal variance v_* with respect to the parameter c turns out to be as follows:

	Value of v_*		
	Portfolio		
Value of c	B	C	H
0.06	0.0000	0.0381	0.0029
.07	.0000	.0232	.0025
.08	.0001	.0132	.0022
.09	.0021	.0081	.0020
.10	.0078	.0080	.0019
.11	.0113	.0129	.0019
.12	.0136	.0227	.0020
.13	.0136	.0375	.0022
0.14	0.0119	0.0372	0.0025
Average 0.10	0.0086	0.0223	0.0022

It is clear that the balanced portfolio H has a much lower minimal variance than that of the growth portfolio B on the average. For a normal population the standard error of the sample variance equals $(2\sigma^4/N)^{1/2}$, where N is the sample size. Hence the following statistic denoting the difference in average v_* values

$$\frac{v_*(B) - v_*(H)}{\sigma(B)} \sim t \qquad (1.4)$$

is distributed as Student's t distribution, where $\sigma(B)$ is the standard error of the average v_* value for portfolio B. The t-value comes to about 2.20 which is clearly statistically significant at the 5% level. For portfolios C and H, the t statistic comes to about 0.01, which is not significant at all. Hence one may conclude that in terms of the average value of v_*, the growth portfolio B differs significantly from the balanced portfolio H, whereas there is no significant difference between the portfolios H and C at the average level of return $c = 0.10$. These results agree with Campanella's findings [6] that showed on the basis of 341 mutual fund portfolios for the 1959-69 decade that the growth funds in Wiesenberger's classification have higher volatility than the balanced funds.

The efficiency frontiers for the three portfolios are

B: $\quad v_* = 0.0016 - 0.0277c + 0.1105c^2$
C: $\quad v_* = 0.2288 - 4.6501c + 24.4301c^2$
H: $\quad v_* = 0.0077 - 0.1121c + 0.5362c^2$

It may be easily checked that there is no real value of c at which any two of the three efficiency frontiers above intersect.

Hence there is no stochastic dominance of any one frontier. However in terms of the value of v_{**} the t-values computed on the basis of the formula (1.4) appear as follows.

	B and H	C and H	B and C
t-value	2.58	2.24	1.34

It is thus clear that the difference in minimal variance v_{**} of the two portfolios B and H is statistically significant. Furthermore, the portfolios B and C are not statistically different by the same test.

3.3 Optimal ordering and selection

In this method we order the sample mean returns from highest to lowest as in Table 3 by using the statistic $\hat{\tau}_j$ for paired comparisons and assert that fund number 5 with a growth objective identifies the best of the ten populations of fund returns. What is the estimate of the true probability of correct selection? The interval estimate is (0.07,0.34) with a point estimate $P_E = 0.205$. Since the point estimate is not greater than $0.55 = 0.50 + .5/k$, k=10 this optimal selection of mutual fund no. 5 is not at an acceptable probability level.

Hence we delete fund no. 5 and rank the remaining nine funds from highest to lowest. In this case (Table 3) we identify fund no. 9 as the best population with the next highest average return. However this has the estimated probability of correct selection $P_E = 0.148$, which is much less than 0.55.

Hence the assertion that fund no. 9 is the best of the nine
funds has an unacceptably low probability of correct selection.
Similar conclusion holds when fund no. 9 is deleted and fund
no. 4 is declared to be the best population. Thus, this
procedure fails to identify the best population with a probability of correct selection exceeding 0.90. We conclude
therefore that in terms of the ordering criterion chosen,
no one fund gives evidence of highest average return with a
high probability of correct selection. Put in another way,
the optimal ordering and selection criteria are not very
sensitive to the data on mean returns for ten randomly selected mutual funds.

3.4 Robustness Tests

The robustness tests proposed here differs from the robust
regression method of Berman, Modiano and Schnabel (1984) in two
ways. First, it is applied to portfolios instead of the individual funds. Second, it maximizes total risk of a given portfolio

$$\max_{x} \sigma^2 = x'Vx \qquad (3.1)$$

$$\text{subject to} \quad x'x = 1; \; x \geq 0$$

subject to a normalization condition. Let $v^* = \max \sigma^2$ be
the maximum value of risk solved from (3.1). We select that
portfolio for which this value v^* of maximum variance is the
least. For example, Table 5 shows very clearly that portfolio
H has the lowest value of v^*, i.e. the minimax eigenvalue.

Portfolio H is robust in the sense of minimax risk. As in the minimax solution of game theory, the selection of portfolio H is the best of the worst in the sense that the highest levels of risk are first considered for each portfolio. Naturally this criterion gives greater weight to pessimistic outcomes (i.e. the worst) but this can be justified when the estimates of mean returns and the covariance of returns are of uncertain precision and quality due to lack of knowledge of the underlying distribution structure.

One notes from the correlation matrix of returns of ten funds (Table 1) that it has all positive elements. Hence the eight portfolios which are linear combinations of ten funds, have themselves a covariance matrix with all positive elements. That is, the matrix V in (3.1) has all its elements positive. Then by Perron-Frobenius theorem, it has a maximal positive eigenvalue λ_1^*, say, associated with which there is an eigenvector x* with positive elements. It follows that

$$v^* = \max \sigma^2 = \lambda_1^*.$$

If the eigenvector x* is normalized so that its elements add up to unity, then one could obtain normalized return for each portfolio by taking the product $\bar{r}'x^*$, where $\bar{r}$ is the vector of mean returns and x* is the normalized eigenvector. Table 5 gives a comparative view of the eight portfolios in respect of the maximum risk levels λ_1^* and the normalized return μ^*, say. For the three reference portfolios B, C and H these are as follows:

Portfolio	Max. Risk (λ_1^*)	Associated Return (μ^*)	Min v_* = $1/\gamma$
B	0.143	0.111	0.0136
C	0.122	0.106	0.0075
H	0.059	0.090	0.0019

It is apparent that the minimax portfolio H which is a balanced portfolio has the lowest maximal risk. It provides the safety-first solution by accepting a lower average return, i.e. about 1.6 to 2.1 percent lower than C or B. Note also that the balanced portfolio H has the lowest level of optimal risk, i.e. minimum v_*. If a t-test is applied in terms of (1.4) for the difference in λ_1^* values for portfolio B and H, it turns out to be statistically significant at 1 percent level (i.e. t=4.12). Thus we may conclude that the balanced portfolio H is most robust.

280 CHAPTER VI

TABLE 1.
Correlation Matrix of Returns of Ten Mutual Funds

	Correlation Coefficients										No. of Row Correlations Exceeding		Mean Return
	1	2	3	4	5	6	7	8	9	10	0.900	0.950	
1	1.000	.937	.949	.907	.909	.874	.952	.956	.935	.969	9	4	.094
2		1.000	.955	.870	.907	.847	.901	.912	.868	.949	7	2	.104
3			1.000	.807	.886	.810	.853	.949	.899	.945	5	1	.088
4				1.000	.696	.769	.936	.858	.846	.821	2	1	.118
5					1.000	.914	.838	.878	.966	.876	5	1	.141
6						1.000	.909	.821	.905	.855	4	1	.101
7							1.000	.936	.971	.920	8	3	.089
8								1.000	.947	.962	7	3	.067
9									1.000	.936	7	3	.123
10										1.000	7	3	.093

TABLE 2.
Various Statistics of the Risk-Return
Efficiency Frontier Equation (1.2)

Portfolio	α	β	γ	k_0	k_1	k_2	$c_* = \beta/\gamma$	$1/\gamma$
1. A	1.131	10.548	91.765	-0.1526	-2.849	-12.39	0.115	0.0109
2. B	1.049	9.239	-73.706	.0016	.028	.1105	.125	.0136
3. C	1.245	12.653	132.913	.2288	4.650	24.43	.095	.0075
4. D	-2.210	-28.339	-359.241	.2315	5.941	37.65	.079	.0028
5. E	4.762	53.201	594.388	36.63	818.43	4,570.8	.089	.0017
6. F	1.319	16.047	205.645	.0960	2.310	14.97	.078	.0049
7. G	-11.470	-72.638	-376.645	.0115	.145	.3766	.193	-.0026
8. H	7.754	56.037	-376.649	0.0077	.112	0.5362	.104	.0019

Note: Portfolios B, C and H emphasize growth, income plus growth and balanced objectives respectively.

TABLE 3.
Probability of Correct Selection of
the Best of Ten Populations of Fund Returns

Objective Code	Fund No.	Mean return (ordered)	$\hat{\delta}_j$	s_j^2	n_j	$\hat{\tau}_j$	P_j
0	5	0.141	—	0.032	16	—	—
0	9	0.123	.018	.032	14	.424	.615
1	4	0.118	.023	.034	15	.543	.647
2	2	0.109	.032	.013	10	.755	.701
0	6	0.101	.040	.018	12	.944	.747
2	1	0.095	.046	.011	14	1.085	.762
2	10	0.093	.048	.013	19	1.132	.786
3	7	0.089	.052	.059	18	1.227	.807
2	3	0.088	.053	.007	15	1.251	.811
2	8	0.067	.074	.015	18	1.747	.892

Note: 1. The last column P_j is the estimated probability of correct selection between two consecutive ordered normal populations.

2. $P_L = P_1 \cdot P_2 \cdots P_{k-1} = 0.07$; $P_U = 0.34$; $0.07 \leq P_E \leq 0.34$.

TABLE 4.
Probability of Correct Selection of the
Best of Nine and Eight Populations of Fund Returns

Fund No. Ordered	Nine Funds (Fund No. 5 Deleted) P_j	Eight Funds (Funds Nos. 5 & 9 Deleted) P_j	Mean Return (Ordered)
9	—		0.123
4	0.510	—	.118
2	0.592	0.561	.109
6	0.641	0.611	.101
1	0.679	0.652	.095
10	0.689	0.664	.093
7	0.714	0.689	.089
3	0.720	0.694	.088
8	0.818	0.807	.067

$P_L = 0.038$ $P_L = 0.057$

$P_U = 0.658$ $P_U = 0.632$

$0.038 \leq P_E \leq 0.658$ $0.057 \leq P_E \leq 0.632$

TABLE 5.
Maximum Eigenvalue of the Variance-Covariance Matrix of Portfolio Returns and the Associated Eigenvector

Portfolio	Maximum Eigenvalues	Associated Eigenvector (row)					Normalized Return
A	0.102	0.338	.530	.543	.557		0.124
B	.143	.233	.445	.729	.464		0.111
C	.122	.283	.249	.491	.786		0.106
D	.108	.300	.251	.848	.357		0.099
E	.143	.257	.752	.531	.294		0.095
F	.067	.395	.318	.746	.431		0.091
G	.071	.315	.721	.451	.421		0.085
H	.059†	.409	.473	.351	.514	(.470)	0.090

Note: 1. Portfolio H has 5 elements in the eigenvector, the last one in parenthesis below 0.514.

2. The last column "normalized return" is the product of mean returns vector and the normalized eigenvector associated with the maximum eigenvalue.

† denotes the minimax eigenvalue

4. Evaluation of Mutual Fund Portfolios

Performance of investment portfolios has three distinct dimensions which can be directly evaluated on the basis of their observed returns and the associated variability. These are for example: (1) the ability of the portfolio manager to increase portfolio returns through successful prediction of future security prices, (2) the ability of the manager to secure efficient diversification of risk that is unsystematic and hence insurable, and (3) the capacity of the manager to provide some robustness in the choice of selected portfolios to investors whose objectives are more for the so-called "balanced" funds than the "growth" funds.

Whereas the fund manager's forecasting ability or the lack of it has been empirically analyzed by a number of researchers, tests of his ability in securing efficient diversification and robustness have not been performed in any detail in respect of empirical data on investment in mutual funds, which are nothing but portfolios with a mixture of securities traded in a stock exchange market. Thus, Jensen [13,16] in his two classic papers analyzed the time-series of returns for 115 mutual funds traded in the New York Stock Exchange over a twenty year period 1945-64 by estimating a regression equation of the form

$$\tilde{R}_{jt} - R_{Ft} = \alpha_j + \beta_j(\tilde{R}_{Mt} - R_{Ft}) + \tilde{u}_{jt} \qquad (1)$$

where $\tilde{R}_{jt}$ = annual continuously compounded rate of return on j-th mutual fund in year t, R_{Ft} = annual continuously compounded risk-free rate of return computed from yield to maturity of a one-year government bond at the beginning of year t, $\tilde{R}_{Mt}$ = estimated annual continuously compounded rate of return on market portfolio calculated from the Standard and Poor Composite 500 Price Index at end of year t and $\tilde{u}_{jt}$ is the error term with zero mean, i.e., $E\tilde{u}_j = 0$. His conclusions were two-fold. The average value of $\hat{\alpha}$, averaged over 115 mutual funds calculated net of expenses was -0.011 implying that on average funds earned about 1.1% less per year. The average value of $\hat{\beta}$ was 0.840 showing that the mutual fund managers tended to hold portfolios that are less risky than the market portfolio $\tilde{R}_{Mt}$. Thus Jensen found little empirical evidence for any individual fund to do significantly better than that which one might expect from mere random chance. Mains [2] relaxed the stationarity assumption behind the systematic risk parameter β_j in (1) but could not reject the inference derived by Jensen that the mutual funds were on average unable to predict security prices well enough to outperform a buy-the-market-and-hold policy. Recently Berman, Modiano and Schnabel [14] made a random selection of ten out of 115 mutual funds from Jensen's data set and reestimated the α_j and β_j coefficients of equation (1) by a robust regression technique, which minimizes the sum of the absolute values of the residuals by a linear programming algorithm. In all cases the least absolute residuals estimates are within one standard deviation of the least squares estimates of Jensen, thus failing to contradict Jensen's finding that the funds are unable to outperform the bechhmarks given by the capital asset pricing model.

The above results of Jensen and Mains have been critically appraised in recent times in two directions. One is Roll's critique [17] that if the market index is ex post efficient in the sense of the mean-variance model, then the performance measure of any mutual fund relative to this market index would not show any abnormal profits on the average. Hence Jensen's result that the fund residuals exhibited no significant departure from the linear CAPM model merely implied that the market index was ex post efficient. The second criticism follows the information-theoretic models of Cornell and Roll [18] and Grossman [19] which predict a market equilibrium where investors who utilize costly information will have higher gross rates of return than their uninformed competitors. But since information is costly, the equilibrium <u>net</u> <u>rates</u> of return for informed and uninformed investors will be the same. This is just what Mains' study reveals, i.e., gross rates of return of mutual funds are greater than the rate on a randomly selected portfolio but the net performance is the same as that for a random investment strategy.

These two criticisms imply that the empirical tests of CAPM must be interpreted with some caution. As noted by Copeland and Weston [1] one could estimate the expected returns directly from the data rather than using a particular model such as the CAPM of asset pricing as a benchmark. Furthermore if investors have heterogeneous expectations then the market portfolio need not be efficient. Since there are different types of mutual funds, e.g., growth fund, balanced fund, etc., one could measure their diversification and robustness characteristics rather than their relative efficiency.

The empirical findings above relate however to only one of three dimensions of performance of investment portfolios. The other two dimensions deal more directly with a portfolio's efficiency in diversification in the Markowitz-Tobin sense and the extent to which the mutual fund portfolios perform a very poor or excellent job of minimizing the "insurable risk" borne by their shareholders. Our objective here is to test the diversification efficiency and robustness of investment portfolios.

4.1 Estimating Equations of the Model

The basic premise of the capital asset pricing model (CAPM) is that since investors hold portfolios of securities rather than a single security, it is most reasonable to measure the riskiness of each security in terms of its contribution to the risk of the portfolio and not its own risk if held in isolation. The subscript j in equation (1) refers however not to any individual security in isolation but a particular portfolio, since a mutual fund is a combination of securities. If only efficient portfolios are considered, i.e., portfolios on the efficiency frontier in the Markowitz-Tobin sense, then the CAPM implies a capital market line as follows:

$$\tilde{R}_p - R_F = \beta_p(\tilde{R}_M - R_F) + \tilde{u}_p \qquad (2)$$

where the subscript p denotes any portfolio on the efficiency frontier and the time subscript is omitted here. If there are m securities in the portfolio p and x_i is the proportion of investment in security i, i=1,2,...,m then it follows

$$\beta_p = \sum_{i=1}^{m} x_i \beta_i; \quad u_p = \sum_{i=1}^{m} x_i u_i \qquad (3)$$

$$\sigma_p^2 = \text{variance of portfolio returns} = \beta_p^2 \sigma_M^2 + \sum_{i=1}^{m} x_i^2 \sigma_i^2$$

where σ_i^2 is the variance of u_i and σ_M^2 that of $\tilde{R}_M$. The systematic risk is given by $(\beta_p^2 \sigma_M^2)$ arising from fluctuations in the market average yield, whereas the second term, $\sum_{i=1}^{m} x_i^2 \sigma_i^2$ represents the unique or unsystematic risk attributable to the individual securities which are in the given portfolio. The greater the diversification of a portfolio the less the unique (i.e., unsystematic) risk of that portfolio. A test of efficiency in diversification must therefore be based on the term $\sum_{i=1}^{m} x_i^2 \sigma_i^2$ and the lowest nonnegative value it can take for a given portfolio.

Tests of robustness of a portfolio may be performed in two ways. One is to analyze the estimated risk-return efficiency frontier in terms of its sensitivity to outlying observations and the second is to apply a minimax-type cautious policy. The first method and for that matter any regression estimate like (1) is not appropriate here, since it is not known if the portfolios are on the efficiency frontier; also the estimates of σ_i^2 which are needed here are not obtainable by the regression methodology, since the optimal set of $x = (x_1, x_2, \ldots, x_m)$ is not directly observable. We adopt therefore a minimax type test of robustness, which is often used in statistical decision theory and the theory of two-person zero-sum games.

To develop this robustness test we proceed as follows. For each portfolio p containing m securities we consider the maximum eigenvalue of the variance-covariance matrix (V) of order m, which gives the maximum of portfolio variance $\sigma^2 = x'Vx$ subject to $x'x = 1$, where x' is the row vector with m elements like x_i. Denote the maximum variance of k-th portfolio by $(\sigma^*(k))^2$. We choose over the set of K portfolios (k=1,2,...,K) by choosing $\min_k (\sigma^*(k))^2$. This defines the minimax portfolio and the associated investment allocation defines the robust or minimax allocation. One difficulty with this minimax allocation is that the elements x_i of the vector x may not all be nonnegative. But we note from Table 1 that the variance-covariance matrix of returns of 10 mutual funds randomly selected by Berman, Modiano and Schnabel [14] from the set of 115 funds analyzed by Jensen has all its elements strictly positive and hence if composite portfolios are constructed by linearly combining the ten funds, the variance-covariance matrix of each such composite portfolio would have strictly positive elements and hence by Frobenius' theorem on positive matrices they would have a strictly positive maximum eigenvalue with a positive eignevector, where the latter could be so normalized that $x'x = 1.0$.

DIVERSIFICATION AND ROBUSTNESS IN PORTFOLIO INVESTMENT

The ten mutual funds out of which we construct composite portfolios are as follows: 1. Composite Fund, Inc. (code 2), 2. Istel Fund, Inc. (code 2), 3. Massachusetts Life Fund (code 2), 4. National Securities - Dividend Series (code 1), 5. Television-Electronics Fund, Inc. (code 0), 6. Group Securities-The Aerospace Science Fund (code 0), 7. Incorporated Investors (code 3), 8. Mutual Investment Fund, Inc. (code 2), 9. United Science Fund (code 0), and 10. The George Putnam Fund of Boston (code 2). Here Weisenberger Financial Services which is the data source lists the fund objectives in codes as 0 = growth, 1 = income, 2 = balanced and 3 = growth-income. We construct the following composite portfolios:

Portfolio	Funds Included	Portfolio	Funds Included
P_1	2,4,5,9	P_5	1,6,7,10
P_2	2,4,6,9	P_6	1,3,7,10
P_3	1,2,4,6	P_7	3,7,8,10
P_4	1,2,6,10	M	1-10
G	5,6,9	B	1-3,8,10
P_{10}	1-4,6-10	P_{11}	1-4,6-8,10
P_{12}	1-3,6-8,10		

Here M is the pseudo-market portfolio since it includes all ten funds randomly selected, B is an all-balanced portfolio, G is an all-growth portfolio and the last three portfolios P_{10} through P_{12} are constructed by deleting each time one fund with the highest mean return hoping that this would reduce portfolio risks accordingly in a sequential way. Since each portfolio involves a number of funds, we could define the efficiency frontier by following the model due to Szego [9] and deriving the optimal allocation vector $x_* = x_*(k)$ and the associated minimal variance for a specific portfolio k, denoted by $\sigma_*^2(k)$. Then one could specify $x_*(k)$ and $\sigma_*^2(k)$ as functions of the parameters $\theta = (V(k), m(k), c(k))$ as follows:

$$x_*(k) = [(\alpha(k) \gamma(k) - \beta^2(k) V(k)]^{-1}$$
$$\cdot [\{m(k) \gamma(k) - e\beta(k)\} c(k) + (e\alpha(k) - m(k) \beta(k))]$$
$$\sigma_*^2(k) = [\alpha(k) \gamma(k) - \beta^2(k)]^{-1} \{\gamma(k) c^2(k) - 2\beta(k) c(k) + \alpha(k)\}$$

where

$$\alpha(k) = m(k)' V^{-1}(k) m(k), \quad \beta(k) = m'(k) V^{-1}(k) e,$$
$$\gamma(k) = e'V^{-1}(k) e$$

On dropping the argument k, the equation for the minimal variance $\sigma_*^2(k)$ may be more simply written as:

$$\sigma_*^2 = ac^2 - bc + d \tag{5}$$

where

$$a = \gamma(k)(\alpha(k) \gamma(k) - \beta^2(k))^1, \quad b = 2\beta(k)(\alpha(k) \gamma(k) - \beta^2(k))^{-1}$$
$$d = \alpha(k)(\alpha(k) \gamma(k) - \beta^2(k))^{-1}$$

The risk-return efficiency frontier (5) is economically meaningful if it is strictly convex and the minimal variance level σ_*^2 never gets negative for values of c within the range $0.06 \leq c \leq 0.14$ which is observed in the market. Again, one could obtain a value of $c_*(k)$ of $c(k)$ at which $\sigma_*^2(k)$ could be further minimized. Thus we get:

$$c_*(k) = \beta(k)/\gamma(k), \quad v_*(k) = \min_c \sigma_*^2(k) = 1/\gamma(k) \ldots \quad (6)$$

which is meaningful if $\beta(k)$ is positive and the optimizing value $c_*(k)$ lies within the range $0.06 \leq c_*(k) \leq 0.14$ observed in the market over the period 1945-64 considered. It is thus clear that when meaningful the composite portfolios P_1 through P_{12} along with M, B and G may be compared for efficient diversification by two statistics: the minimal variance $\sigma_*^2(k)$ and the lowest point $v_*(k)$ of the risk-return efficiency frontier $\sigma_*^2(k) = f(c(k))$ viewed as a strict convex function of $c(k)$, provided $v_*(k)$ is nonnegtive.

4.2 Econometric Tests of Diversification

The statistical tests of diversification applied here are intended to detect for different portfolios, whether their minimal risk levels differ in a statistically significant sense. A reference portfolio is selected, e.g., the pseudo-market portfolio M in this case and all other portfolios are compared in respect of the reference portfolio for their estimated mean returns and variances. In particular, three portfolios, e.g., balanced portfolio B, growth portfolio G and a typical mixed portfolio P_3 are compared in respect of portfolio M.

The statistics used to compare the portfolios reported in Table 2 are based on the risk-return efficiency frontier equation

(5): $\sigma_*^2 = f(c)$, which specifies minimal risk (i.e., minimum variance) for a given c within the observed range $0.06 \leq c \leq 0.14$. Since the expected return parameter c is inclusive of brokerage fees and expenses of management of funds and also the risk-free rate of return, it appears more meaningful to consider the realistic range of c to be $0.09 \leq c \leq 0.14$. Values of minimal risk for selected portfolios on the risk-return efficiency frontier equation (5) turn out to be as follows:

Minimal Risk (σ_*^2) of Portfolios

Value of c	B	M	G	P_3
0.09	0.00197	0.0046	NM	0.0082
.10	.00187	.0048	NM	.00809
.11	.00188	.0050	0.0042	.0129
.12	.00199	.0052	.0073	.0227
.13	.0022	.0055	.0095	.0374
0.14	0.0025	0.0146	0.0106	0.0572

NM denotes a number which is negative or very close to zero and hence not meaningful.

Several interesting points emerge from this table. First, the minimal risk level for the balanced portfolio B is the lowest for any level of the expected return parameter c; hence it dominates the other portfolios M, G, and P_3. Second, the typical mixed portfolio P_3 has higher minimal risk than the market portfolio M for any level of c. Moreover the difference $\delta = \sigma_*^2(P_3) - \sigma_*^2(M)$ between the two portfolios' minimal risk levels increases as c rises from 0.10 to 0.14. Third, the two portfolios M and G intersect each other at a value of c = 0.113, which implies that the investor would remain indifferent between M and G in respect of diversification efficiency, if his expected return c is between 11% and 12%.

The difference in minimal risks for the four portfolios may be statistically tested by a t-statistic, if the return variable $\tilde{R}_{jt}$ is lognormally distributed. In Jensen's data set the return variable $\tilde{R}_{jt}$ is indeed lognormally distributed, since it is measured in logarithmic units and the linear regression model (1) assumes normality. Hence the standard error of the estimated quantity $\hat{\sigma}_*^2$ is given by $(\sigma^2*(2/N)^{\frac{1}{2}})$ where σ^2* is the population variance. Hence for any two portfolios, e.g., B and M one could test the difference in minimal risk levels by

$$t = (N/2)^{\frac{1}{2}} \frac{\hat{\sigma}_*^2(M) - \hat{\sigma}_*^2(B)}{[\hat{\sigma}_*^4(M) + \hat{\sigma}_*^4(B)]^{\frac{1}{2}}} \qquad (7)$$

where N is the sample size. On using this statistic (7), the computed t values for the four different portfolios B, M, G and P_3 are as follows:

	Values of t for Portfolios		
Value of c	M & B	G & B	P_3 & M
0.11	1.76*	2.15**	1.61
0.12	1.71*	2.18**	2.25**
0.13	1.68	2.26**	2.54**
0.14	2.45**	2.23**	2.17**

Note: One and two asterisks denote a significant t at 10% and 5% levels respectively.

It is clear that the balanced portfolio is significantly different from the growth portfolio at the 5% level of one-sided t-test and at 10% level the balanced portfolio has a significantly lower minimal risk than the market portfolio. Furthermore for expected returns on or above 12%, the market portfolio M beats P_3 in a statistically significant way at the 5% level of t-test.

In a similar way one could use the ttest (7) for testing the difference in the four portfolios in terms of their least optimal risk levels $v_*(k)$ defined in (6). The computed values of t turn out to be as follows:

	M & B	G & B	P_3 & M
t-value	1.70*	2.43**	0.70

The balanced portfolio turns out to be significantly different from the other two portfolios, although the market portfolio M and the random portfolio P_3 are not so different at all in a statistically significant sense. The latter result presents a striking contrast to the earlier comparison of the minimal risk levels σ_*^2, where P_3 and M differed significantly at the 5% level of t-test. The reason for this contrast is not far to seek. The comparison of $v_*(k)$ is associated with different levels of $c_*(k)$. Hence a more appropriate comparison would be a test of the difference of values of $c_*(k)$ in terms of the following t-statistic:

$$\frac{\sqrt{N}|c_*(M) - c_*(B)|}{\sqrt{\{v_*(M) + v_*(B)\}}} \sim |t|$$

The computed values of $|t|$ turn out to be:

	M & B	G & B	P_3 & M
	5.53***	14.08***	30.15***

It is clear that the difference between the portfolio-pairs is highly significant at even the 1% level of $|t|$ test, when expected return levels are compared. Note that three asterisks here denote significant $|t|$ value at 1% level.

For robustness we apply the maximum eigenvalue criterion, where the maximum eigenvalue (λ_1^*) of the k-th portfolio is defined by

$$(V(k) - \lambda I) x(k) = 0 \qquad (8.1)$$
$$x'(k) x(k) = 1.0 \qquad (8.2)$$

where (8.2) defines the normalization of the eigenvector $x(k)$ and by Frobenius' theorem the maximum eigenvalue λ_1^* which is positive satisfies (8.1) and (8.2) as:

$$(V(k) - \lambda_1^* I) x^*(k) = 0; \quad x^{*\prime}(k) x^*(k) = 1$$

with the elements of the eigenvector $x^*(k)$ being positive. By premultiplying (8.1) by $x^*(k)$ and applying the normalization condition (8.2) we obtain

$$\lambda_1^* = x^{*\prime}(k) V(k) x^*(k) = (\sigma^*(k))^2$$

Hence the difference of two Frobenius' eigenvalues, $\lambda_1^*(G)$ and $\lambda_1^*(B)$ say may be statistically tested by a t-statistic as in (7), i.e.,

$$\frac{(N/2)^{\frac{1}{2}} (|\lambda_1^*(G) - \lambda_1^*(B)|)}{[\lambda_1^{*2}(G) + \lambda_1^{*2}(B)]^{\frac{1}{2}}} \sim |t| \qquad (9)$$

CHAPTER VI

The results of this tests applied to the maximum eigenvalues (λ_1^*) given in Table 3 are as follows:

	B & G	M & B	P_3 & B	P_3 & M	P_3 & P_1
$\|t\|$-value	16.36***	22.70***	14.54***	14.48***	3.77***

It is clear that the portfolio-pairs are significantly different at 1% level of $|t|$-test. Of all the portfolios considered here, the balanced portfolio is the most robust, since it has the lowest value of the Frobenius' eigenvalue λ_1^*, which has the normalized mean return of 9%. Since $\lambda_1^*(B)$ is the minimax eigenvalue, it is much less sensitive to departures from the assumption of normality that is implicit in the $|t|$-statistic defined in (9). Also, for undersized samples (i.e., when the sample size N is too small to estimate with any precision the standard error) the minimax eigenvalue $\lambda_1^*(B)$ can be applied in a game-theoretic sense as a safety-first or cautious decision rule.

Besides the comparison of the balanced portfolio with the others, a second interesting comparison set is provided by the four portfolios: M, P_{10}, P_{11}, P_{12}, where each succeeding portfolio is obtained by deleting a mutual fund with a highest mean return. The relative decline in λ_1^*-values shows how sensitive is the maximal risk to decreases in normalized mean returns. The computed $|t|$-values defined as in (9) appear as follows:

	M & P_{10}	P_{10} & P_{11}	P_{11} & P_{12}
$\|t\|$:	2.59**	3.05***	3.23***

Note that the normalized mean returns are 10.4%, 10.0%, 9.6% and 9.3% for the respective portfolios M, P_{10} through P_{12}. The minimax portfolio in this case is P_{12} which is significantly different from P_{11} and hence the others at 1% level of the |t|-test. It is thus clear that the robustness in choice between portfolios or mutual funds may be pursued by the investor as a desirable policy of caution, when the parameter estimates of the probability distribution of returns may not be obtained with a high degree of precision.

The eigenvalue statistics of Table 3 show another striking result most clearly. The Frobenius' eigenvalue of the market portfolio M is the highest of all the portfolios. This has two important implications. One is that any portfolio P_j (j=1,2,...,7,10,11,12) is less risky than the market portfolio. This corresponds very closely with Jensen's finding through the β_j coefficient defined in (1) that most fund managers tend to hold portfolios which were less risky than the market portfolio. The second implication is that this establishes a very strong case for selecting a balanced portfolio for most risk-averse investors. The trade-off between risk-reduction and the reduction in expected return may be substantial and also statistically significant.

The above results have two significant policy implications. One is that the portfolios of mutual funds, when suitably chosen may be doing an excellent job of diversification in its two facets: risk-minimization and robustness. This may itself be a significantly socially desirable service to investors.

CHAPTER VI

Second, this analysis casts some doubt on the hypothesis of capital market efficiency in its semi strong form, which through the CAPM model suggests that the portfolio and fund managers can earn abnormal profits through successful prediction of future security prices.

Table 1. Mean, Variance and Some Covariance Terms of Ten Randomly Selected Mutual Fund Returns

Fund No.	Mean	Variance	Covariance	
			Maximum	Minimum
1	0.094	0.010	0.027	0.008
2	0.104	0.013	0.022	0.009
3	0.088	0.007	0.022	0.008
4	0.118	0.034	0.043	0.012
5	0.141	0.032	0.051	0.014
6	0.101	0.080	0.059	0.013
7	0.089	0.038	0.059	0.015
8	0.067	0.015	0.034	0.010
9	0.123	0.032	0.049	0.014
10	0.093	0.013	0.032	0.010
Average	0.102	0.027		

Table 2. Selected Statistics of Diversification of Composite Portfolios

Portfolio	Equation (5) Parameters			$c_*(k)$	$v_*(k)=1/\gamma(k)$	CV
	a	b	d			
P_1	-12.281	-	-	NM	NM	
P_2	0.110	0.028	0.002	0.125	0.0136	
P_3	24.748	4.706	0.232	0.095	0.0075	0.911
P_4	38.651	6.098	0.238	0.079	0.0028	
P_5	1408.410	251.104	53.744	0.089	0.0017	
P_6	14.952	2.333	0.096	0.078	0.0049	
P_7	0.394	0.145	0.012	0.193	NM	
B	0.527	0.112	0.008	0.104	0.0019	0.419
M	0.243	0.031	0.006	0.066	0.005	1.071
G	-5.007	-1.464	-0.096	0.146	0.011	0.718
P_{10}	1.117	0.101	0.007	0.045	0.005	
P_{11}	0.543	0.137	0.006	0.131	NM	
P_{12}	0.404	0.074	-0.002	0.092	NM	

Note: (a) NM = not meaningful
(b) B = balanced, M = pseudo market portfolio, G = growth
(c) CV = coefficient of variations = $\sqrt{v_*(k)}/c_*(k)$

Table 3. Maximal Frobenius' Eigenvalues (λ_1^*) of Equation (7) and
Normalized Mean Returns of Portfolios

	Portfolio						
	P_1	P_2	P_3	P_4	P_5	P_6	P_7
Maximal Eigenvalue λ_1^*	0.102	0.143	0.122	0.108	0.143	0.067	0.071
Minimum Positive Eigenvalue	0.002	0.004	0.002	0.010	0.0001	0.0002	0.0002
Normalized Mean Return	0.124	0.111	0.106	0.099	0.095	0.091	0.085
	B	M	G	P10	P_{11}	P_{12}	
Maximal Eigenvalue λ_1^*	0.059	0.261	0.143	0.231	0.200	0.172	
Minimum Positive Eigenvalue	0.0001	0.0001	0.002	0.0002	0.0002	0.0001	
Normalized Mean Return	0.090	0.104	0.118	0.100	0.096	0.093	

Note: 'Normalized mean return' is the inner product of the mean return vector and the normalized eigenvector associated with the maximal eigenvalue; the eigenvector is defined to be normalized, if the sum of squares of its elements is unity.

5. Estimation of Portfolio Efficiency Frontier

Investment portfolios are often analyzed in financial management literature to test the performance of portfolio managers such as money market funds. The rapid growth of financial assets controlled and managed by institutional investors such as pension and retirement funds and open-end trusts has generated a demand for realistic standards for judging portfolio performance: (a) the ability of the manager to minimize risk through diversification and to predict and take advantage of general market turns, and (b) his performance in maintaining a portfolio consistent with a stated investment objective with respect to growth of capital, income characteristics and liquidity.

The prediction of general market turns, i.e., upswings and downswings may be tested by comparing it with the realized values at the beginning of the turning points. Campanella [6] made such a test on the basis of 339 mutual fund portfolios for the decade 1959-69. Using $B_t = \sum_{j=1}^{n} \beta_j x_j t$ as the measure of systematic portfolio risk at time t, which is the weighted sum of the systematic risk elements (β_j) of its component securities, he plotted for each mutual fund quarterly values of B_t along with the quarterly levels of the Standard and Poor's 500 Composite Index. If a fund is successful in adjusting portfolio volatility (here measured by B_t) in anticipation of market movements, plotted volatilities should appear as a leading series, since an anticipated market downturn (upturn) would be preceded by a decrease (increase) in portfolio volatility. However this is not the case in more than 90% of the cases. The following estimates by Campanella show this very clearly:

Quarterly Volatility Levels (B_t) at t Related to
Market Movements at t+1

	Market Up			Market Down		
	Above Average	Below Average	Average	Above Average	Below Average	Average
No. of Funds	94	121	6	74	38	6

In general the growth funds with the objective of maximum capital gain fared poorer than the balanced funds.

The second test of investment performance is to assess the capacity of the portfolio manager to provide some robustness in optimal portfolio choice, when significant estimation risks exist.

Three types of problems arise when one attempts to estimate the mean-variance efficiency frontier against observed empirical data on mutual fund returns, treating each fund as a composite portfolio. First, the mean variance parameter set (m,V) is not directly observable from market information. Also, a direct estimate of variance-covariance matrix V is out of the question due to the requirement of large data sets, which are unavailable. The estimation risk plays a critical role here. Second, the optimal vector x_* of investment proportions is usually not available from published time-series data and hence it is not possible to apply the test statistic proposed by Roll [7]: $z=\hat{x}'_*\hat{x}_*$ which is asymptotically normal with mean x'_*x_* and variance whose sample analog is $x'_*(\hat{x}_*\hat{x}'_*)x_*/N$. Here $\hat{x}_*$ denotes the vector of investment proportions for the sample efficient portfolio, where $y=\tilde{r}'x$ is assumed to be normally distributed with mean m'x and variance x'Vx. Third, a Bayesian approach with a diffuse prior leads to the same optimal allocation vector as x_*, so that any sequential improvement with more information is not possible. As Bawa, Brown and Klein [4] have shown that if prior to any observation the portfolio manager is ignorant of the values of the parameters, then he may derive the predictive distribution of portfolio returns, which follows Student's t-distribution with mean $\mu^0 = \hat{\mu} = (\sum_{t=1}^{N} y_t)/N$ and variance $\sigma^{02} = k_N \hat{\sigma}^2 = k_N (\sum_{t=1}^{N} (y_t - \hat{\mu})^2)/(N-1)$, $k_N = (1+1/N)(N-1)/(N-3)$ with (N-1) degrees of freedom. It is clear that $V^0 = k_N \hat{V}$, where $k_N = (1+\frac{1}{N})(N-1)(N-n-2)$ and $\hat{V}$ is the usual sample estimate of the covariance matrix of returns for a given sample of size N where n is the number of risky assets. Clearly any portfolio x which minimizes the estimated portfolio variance $\hat{\sigma}^2 = x'\hat{V}x$ will also minimize the predictive variance

$$\sigma^{02} = x'V^0 x = k_N x'\hat{V}x = k_N \hat{\sigma}^2$$

for a given mean return $\mu^0{}'x = \hat{\mu}'x = c$, where $\mu^0, \hat{\mu}$ are estimates of the mean vector m. In other words, the ex ante mean-variance efficiency is not distinguishable from its expost form, when the observations consist of (x_*, σ_*^2) only.

We have adopted therefore a direct regression test of the efficiency frontier equation, as if each mutual fund investment decision were in fact an optimal portfolio decision, given the market information available at that time. For estimation purposes however, three specifications of the efficiency frontier equation may be suggested as follows:

$$(R_{jt}-R_{Ft})^2 = \alpha_j + \beta_j(R_{Mt}-R_{Ft}) + \gamma_j(R_{Mt}-R_{Ft})^2 + u_{jt} \quad (1.1)$$

$$R_{jt}-R_{Ft} = \alpha_j + \beta_j(R_{Mt}-R_{Ft}) + u_{jt} \quad (1.2)$$

$$R_{jt}^2 = \alpha_j + \beta_j R_{Mt} + \gamma_j R_{Mt}^2 + u_{jt} \quad (1.3)$$

where R_{jt} = annual continuously compounded rate of return on j-th mutual fund in year t, R_{Ft} = annual continuously compounded risk-free rate of return computed from yield to maturity of a one-year government bond at the beginning of year t, R_{Mt} = estimated annual continuously compounded rate of return on market portfolio calculated from the Standard and Poor Composite 500 Price Index at end of year t and u_{jt} is the random disturbance term with zero mean and fixed variance for all t.

Assuming R_{Ft} to be nonstochastic, the specification (1.1) leads to the following form of conditional variance of R_{jt}, given $(R_{Mt}-R_{Ft})$:

$$\text{Var } R_j = a_j + \beta_j(R_{Mt}-R_{Ft}) + \gamma_t(R_M-R_{Ft})^2 \quad (2.1)$$

which is analogous to the frontier equation, where $c \simeq (R_{Mt}-R_{Ft})$, $a_j = \alpha_j - (\bar{R}_j - R_{Ft})^2$, $\bar{R}_j = E(R_{jt})$ and $v = \text{Var } R_{jt} = E[(R_{jt}-\bar{R}_j)^2]$. The second specification (1.2) follows the capital asset pricing model (CAPM), which requires that in equilibrium, the market portfolio with return R_{Mt} must be an efficient portfolio in the mean-variance sense and the equilibrium rates of return on all risky assets are a function of their covariance with the market portfolio. This implies that the coefficient β_j in (1.2) may be interpreted as

$$\beta_j = \text{Cov}(R_{jt}, R_{Mt})/\text{Var}(R_{Mt}) \tag{2.2}$$

If all investors have homogeneous expectations and the bivariate distributions of R_{jt} and R_{Mt} are approximately normal, then the market portfolio can be shown to be efficient and in that case the mean-variance efficiency in ex ante form can be measured by the specification (1.2), with the regression coefficient $\hat{\beta}_j$ defined in (2.2) estimated by the ex post time-series data.

The third specification (1.3) uses a typical market model but in a slightly different quadratic form. For instance, a typical market model is written as

$$\tilde{R}_j = a_j + b_j \tilde{R}_M + \tilde{\varepsilon}_j \tag{3.1}$$

where the random return on any mutual fund or composite asset j, ($\tilde{R}_j$) is viewed as a linear function of market return ($\tilde{R}_M$) plus a random error term $\tilde{\varepsilon}_j$, which is assumed to be independent of the market return. Total risk of fund j measured by its variance may then be partitioned as:

$$\text{Var}(\tilde{R}_j) = b_j^2 \text{Var}(\tilde{R}_M) + \sigma_\varepsilon^2 \tag{3.2}$$

where $b_j^2 \text{Var}(\tilde{R}_M)$ is systematic risk and σ_ε^2 is unsystematic risk. Instead of the linear form (4.1), the specification (2.3) uses R_{jt}^2 as the dependent variable and a quadratic form, which leads to the following variance function

$$\text{Var } R_j = \delta_j + \beta_j \bar{R}_M + \gamma_j \text{Var } R_M \tag{3.3}$$

where

$$\delta_j = \alpha_j - \bar{R}_j^2 + \gamma_j \bar{R}_M^2, \quad \bar{R}_j = E(R_{jt}), \quad \bar{R}_M = E(R_{Mt})$$

This variance function (2.3), closely analogous to the earlier form (2.1) does not incorporate the variable R_{Ft} and the variance term σ_u^2; hence it may not satisfy the full requirements of the standard CAPM of market efficiency e.g. the presence of heterogeneous expectations by investors and their weights on the skewness of the nonnormal return distributions may be better captured by this specification, if it is empirically validated by ex post time series data.

Two aspects of estimation risk may be analyzed through the empirical regression estimates of the three specifications (1.1) through (1.3) as above. First, one may decide in terms of empirical fit (i.e. R^2 and t values of regression coefficients) which of the three models perform the best in terms of correspondence with observed data. Second, one may test if there is any significant difference between different portfolios or, mutual funds in terms of a suitable robustness criterion. For example, Levy [20] found by empirical estimates that the portfolios which are diversified by industry groups tend to perform better than the portfolios which are diversified randomly.

5.1 Empirical Estimation and Implications

For empirical estimation of the three specifications of the portfolio efficiency frontier, we utilized the data sets on returns for ten mutual funds, which were previously selected randomly by Berman, Modiano and Schnabel out of a larger set of 115 funds used in Jensen's study mentioned earlier. These ten mutual funds had their specific portfolio objectives listed in codes by Weisenberger Financial Services as follows: 0 for growth, 1 for income, 2 for balanced and 3 for for growth plus income. Thus growth funds or portfolios emphasized growth of capital much more than income or balanced funds, whereas balanced funds seem to place more weight on the objective of risk-aversion. The ten funds for which the available yearly data over 1945-64 were used in our empirical study are as follows: 1. Composite Fund, Inc. (code 2), 2. Istel Fund, Inc. (code 2), 3. Massachusetts Life Fund (code 2), 4. National Securities-Devidend Series (code 1), 5. Television-Electronics Fund, Inc. (code 0), 6. Group-Securities: The Aerospace Science Fund (code 0), 7. Incorporated Investors (code 3), 8. Mutual Investment Fund, Inc. (code 2), 9. United Science Fund (code 0), and 10. The George Putnam Fund of Boston (code 2). The mean, variance and some covariance terms of returns on these funds are as follows:

Fund	Code	Mean	Variance	Covariance Max	Min
1	2	0.094	0.010	0.027	0.008
2	2	.109	.013	.022	.009
3	2	.088	.007	.022	.008
4	1	.118	.034	.043	.012
5	0	.141	.032	.051	.014
6	0	.101	.080	.059	.013
7	3	.089	.038	.059	.015
8	2	.067	.015	.034	.010
9	0	.123	.032	.049	.014
10	2	0.093	0.013	0.032	0.010

From our earlier analysis we found that the correlation matrix of returns of these funds has all positive elements, with a minimum correlation of 0.769 and a maximum off-diagonal coefficient of 0.971. This shows that as portfolios the different funds have a great degree of similarity in diversification. Moreover the growth funds (code 0) significantly differ in a statistical sense from the balanced funds (code 2) in two distinct ways. First, they have much higher variance of returns and a higher spread between the lowest and highest variance. Second, the growth funds are more strongly correlated among themselves than with the balanced funds. This suggests that we can test if there is any difference in performance of the growth and balanced funds in terms of our estimation of the efficiency frontier equations (1.1) through (1.3), which are reported in Tables 1-4 at the end.

From an overall point of view the CAPM specification (1.2) in its linear form comes out to be the best when measured by R^2 and the statistical significance of the relevant regression coefficients. The R^2 values have the following ranges $0.67 \leq R^2 \leq 0.93$ for (1.1), $0.74 \leq R^2 \leq 0.95$ for (1.2) and $0.71 \leq R^2 \leq 0.93$. However, if fund no. 4 which is an income fund is set aside, the range for (1.1) improves to $0.74 \leq R^2 \leq 0.93$, while it remains the same for (1.2) and (1.3). This shows very clearly that the quadratic efficiency frontier (1.1) which follows directly from the variance minimization model, can be a good substitute for the linear risk-return relationship (1.2) derived from the CAPM of market efficiency. Two other characteristics of the quadratic efficency frontier (1.1) need to be emphasized. First, it does not require the bivariate normality of R_{jt} and R_{Mt} as in (1.2) and hence the slope coefficient γ_j directly measures the risk sensitivity of j-th fund i.e., $\gamma_j = \partial(\text{Var } R_j)/\partial(\text{Var } R_M)$. In terms of this risk sensitivity measure the growth funds are far more risk-sensitive than the balanced funds as follows:

	income fund	growth fund			balanced funds			
	4	5	6	9	1	2	3	8	10
value of $\hat{\gamma}_j$	1.29	1.68	4.69	1.43	0.48	0.83	0.37	0.57	0.67

On taking three typical funds one from each group and performing t-tests on the difference of $\hat{\gamma}_j$ values we find that balanced funds differ significantly from growth funds at 1% level of t-test and from the income fund at 5% level.

values of t		
Funds	Funds	Funds
1 & 9	1 & 4	9 & 4
3.06**	2.61*	0.39

Secondly, one may easily test if for any two funds 1 and 9 say, the two quadratic frontier functions intersect at any positive value of $c = R_M - R_F$ within the realistic range $0.06 \leq c \leq 0.14$. If they intersect at a value say c_0, then one portfolio may dominate the other for a value of $c > c_0$. This follows from the theory of stochastic dominance and fund separability properties. On equating the two frontier functions for the balanced fund 1 and growth fund 9, one finds the intersection point $c_0 = 0.09$, to the right of which the balanced fund dominates the growth fund in the sense $y_1 < y_9$ where $y_j = E(R_{jt}-R_{Ft})^2$. It is clear therefore that in terms of efficiency in diversification broadly conceived, there is a significant difference between the two groups of funds, the growth and balanced funds.

For comparative purposes we also performed tests of nonnormality of the residuals from the linear and quadratic regression of $R_j - R_F$ on $R_M - R_F$ as specified in (1.2) and (1.1). The Shapiro-Wilk test statistic is used, which specifies that for normality of the residuals the calculated value of W should be larger than the tabulated value denoted by W_0 at 50%. The results are as follows:

Fund	Code	Linear			Quadratic		
		W	W_0	Remark	W	W_0	Remark
1	2	0.9374	0.947	NN	0.9346	0.947	NN
2	2	0.9415	0.938	-	0.9241	0.938	NN
3	2	0.9226	0.950	NN	0.8878	0.950	NN
8	2	0.9790	0.956	-	0.9078	0.956	NN
10	2	0.9708	0.957	-	0.9551	0.957	NN
5	0	0.9697	0.952	-	0.9570	0.952	-
6	0	0.9061	0.943	NN	0.8196	0.943	NN
9	0	0.9549	0.947	-	0.9718	0.947	-
4	1	0.9867	0.950	-	0.9503	0.950	-

Note: NN stands for 'nonnormal'.

It is clear that the quadratic model has more nonnormal residuals and this is more so for the balanced funds. For example all the five balanced funds show evidence of nonnormality of residuals in the quadratic regression model, as against only two in the linear model. Since the R^2 values for the linear and quadratic models are very close, this leads to the implication that the specification of the quadratic model could be improved so as to allow for the nonnormality of distribution of fund returns. We could also interpret the result slightly differently as follows: If the underlying population distribution of fund returns ($R_j - R_F$) is nonnormal, then the estimation of the quadratic regression model shows it very clearly but within the limits of Shapiro-Wilk test.

6. Concluding Remarks

Our empirical analysis has shown that the issue of diversification in optimal portfolio investment has several facets, when realistic common stock and mutual fund data are considered with mean variance parameters estimated. The concept of robustness in some sense and its empirical applications to measure portfolio efficiency open up a new potential field of research which may prove helpful to most investors.

Table 1. Quadratic Regression of $(R_j-R_F)^2$ on (R_M-R_F) for Ten Mutual Funds

Fund	Code	Intercept	(R_M-R_F)	$(R_M-R_F)^2$	R^2	N
1	2	0.00002 (.0076)	-0.0296 (-1.66)	0.4776** (7.47)	0.91	14
2	2	-0.0053 (-1.07)	-0.0172 (-0.48)	0.8301** (4.88)	0.87	10
3	2	-0.0019 (-0.75)	-0.0112 (-0.54)	0.3712** (5.07)	0.84	15
4	1	0.0098 (0.92)	-0.1478 (-1.70)	1.2951** (4.10)	0.67	15
5	0	-0.0054 (-0.42)	-0.1132 (-1.08)	1.6843** (4.43)	0.74	16
6	0	-0.0312 (-0.85)	-0.6672* (-2.21)	4.6954** (4.37)	0.74	12
7	3	0.0045 (1.01)	-0.1988** (-5.11)	1.7055** (11.86)	0.93	18
8	2	0.0036 (1.36)	-0.0590* (-2.54)	0.5670** (6.60)	0.82	18
9	0	0.0003 (0.0457)	-0.1189* (-2.42)	1.4280** (8.12)	0.91	14
10	2	-0.0012 (-0.3701)	-0.0344 (-1.52)	0.6749** (8.14)	0.90	19

Note: 1. t-values are in parenthesis, one and two asterisks denote significance at 5 and 1% respectively.

2. Codes for mutual funds are: 0 for growth, 1 for income, 2 for balanced and 3 for growth and income.

CHAPTER VI

Table 2. Linear Regression of $(R_j - R_F)$ on $(R_M - R_F)$ for Ten Mutual Funds

Fund	Code	Intercept	$(R_M - R_F)$	R^2	N
1	2	-0.0013 (-0.0132)	0.6108** (12.23)	0.93	14
2	2	-0.0172 (-1.083)	0.6926** (7.65)	0.88	10
3	2	-0.0021 (-0.199)	0.5231** (9.88)	0.88	15
4	1	-0.0427 (-1.720)	1.0976** (8.77)	0.86	15
5	0	-0.0112 (-0.369)	1.0229** (6.59)	0.74	16
6	0	-0.0747 (-1.418)	1.4161** (5.29)	0.74	12
7	3	-0.0616** (-4.643)	1.2359** (17.18)	0.95	18
8	2	-0.0347** (-3.663)	0.7700** (15.00)	0.93	18
9	0	-0.0250 (-1.262)	1.0663** (10.40)	0.90	14
10	2	-0.0045 (-0.420)	0.7186** (12.15)	0.90	19
Average		0.0275	0.9154		

Note: 1. t-values are in parenthesis, one and two asterisks denote significance at 5 and 1% respectively.

2. Codes for mutual funds: 0 = growth, 1 = income, 2 = balanced and 3 = for growth and income.

Table 3. Linear Regression of R_j on R_M for Ten Mutual Funds

Fund	Code	Intercept	(R_M-R_F)	R^2	N
1	2	0.0098 (0.85)	0.6048** (10.85)	0.91	14
2	2	0.0027 (1.46)	0.6915** (7.15)	0.86	10
3	2	0.0117 (0.94)	0.5116** (8.68)	0.85	15
4	1	-0.0457 (-1.67)	1.1012** (8.48)	0.85	15
5	0	-0.0132 (-0.39)	1.0323** (6.42)	0.75	16
6	0	-0.0866 (-1.49)	1.4182** (5.12)	0.72	12
7	3	-0.0668** (-4.64)	1.2381** (16.77)	0.95	18
8	2	-0.0297** (-2.85)	0.7694** (14.42)	0.93	18
9	0	-0.0272 (-1.25)	1.0697** (10.11)	0.89	14
10	2	0.0014 (0.12)	0.7207** (11.57)	0.88	19

Note: One and two asterisks denote 5 and 1% levels of signifiance of t-values given in parenthesis.

CHAPTER VI

Table 4. Linear and Quadratic Regression of R_j^2 on R_M and R_M^2 for ten Mutual Funds

Fund	Intercept	R_M	R_M^2	R^2	N
1	0.0031	0.1081**	-	0.61	14
	0.0013	-0.0174	0.4534**	0.93	
2	0.0052	0.1548**	-	0.54	10
	-0.0033	-0.0338	0.8729**	0.87	
3	-0.00005	0.0985**	-	0.60	15
	-0.0015	-0.0094	0.3928**	0.84	
4	0.0162	0.2009*	-	0.36	15
	0.0114	-0.1514	1.2835**	0.72	
5	0.0022	0.3201**	-	0.42	16
	-0.0022	-0.1614	1.7576**	0.75	
6	0.0179	0.4934	-	0.22	12
	-0.0123	-0.8682*	4.8041**	0.71	
7	0.0143	0.2346*	-	0.38	18
	0.0068	-0.2143**	1.6837**	0.92	
8	0.0058	0.1022**	-	0.49	18
	0.0033	-0.0481	0.5653**	0.84	
9	0.0081	0.2603**	-	0.51	14
	0.0026	-0.1416*	1.4469**	0.91	
10	0.0022	0.1472**	-	0.58	19
	-0.0001	-0.0341	0.6831	0.90	

Note: One and two asterisks denote significant t-values at 5 and 1% levels respectively.

References

1. Copeland, T.E. and J.F. Weston. Financial Theory and Corporate Policy. Reading, Massachusetts: Addison-Wesley, 1983.
2. Mains, N.E. Risk, The Pricing of Capital Assets and the Evaluation of Investment Portfolios: Comment. Journal of Business 50 (1977), 371-384.
3. Beedles, W.L. On the Asymmetry of Market Returns. Journal of Financial and Quantitative Economics 14 (1979), 653-660.
4. Bawa, V.S., Brown, S.J. and R.W. Klein. Estimation Risk and Optimal Portfolio Choice. Amsterdam: North Holland, 1979.
5. Arditti, F.D. and H. Levy. Portfolio Efficiency Analysis in Three Moments: The Multiperiod Case, in H. Levy and M. Sarnat, eds. Financial Decision Making Under Uncertainty. New York: Academic Press, 1977.
6. Campanella, F.B. The Measurement of Portfolio Risk Exposure: Use of the Beta Coefficient. Lexington, Massachusetts: D.C. Heath, 1972.
7. Roll, R. Testing a Portfolio for Ex-ante Mean-variance Efficiency, in E.J. Elton and M.J. Gruber, eds. Portfolio Theory: Twenty Five Years Later. Amsterdam: North Holland, 1979.
8. Jacob, N.L. A Limited Portfolio Selection Model for the Small Investor. Journal of Finance 29 (1947), 847-856.
9. Szego, G.P. Portfolio Theory. New York: Academic Press, 1980.

10. Sengupta, J.K. A Minimax Policy for Optimal Portfolio Choice. International Journal of Systems Science 13 (1982), 39-56.
11. Toutenberg, H. Prior Information in Linear Models. New York: John Wiley, 1982.
12. Tintner, G. Multicollinearity. Working Paper of the Institute of Econometrics. Vienna, Austria, 1979.
13. Jensen, M.C. The Performance of Mutual Funds in the Period 1945-64. Journal of Finance 23 (1968), 389-416.
14. Berman, O., Modiano, E. and J.A. Schnabel. Sensitivity Analysis and Robust Regression in Investment Performance Evaluation. International Journal of Systems Science 15 (1984), 475-480.
15. Gibbons, J.D., Olkin, I. and M. Sobel. Selecting and Ordering Populations: A New Statistical Methodology. New York: John Wiley, 1977.
16. Jensen, M.C. Risk, the Pricing of Capital Assets and the Evaluation of Investment Portfolios. Journal of Business 42 (1969), 167-247.
17. Roll, R. A Critique of the Asset Pricing Theory's Tests. Journal of Financial Economics 6 (1977), 129-176.
18. Cornell, B. and R. Roll. Strategies for Pairwise Competitions in Markets and Organizations. Bell Journal of Economics 8 (1981), 201-213.
19. Grossman, S. The Impossibility of Informationally Efficient Markets. American Economic Review 21 (1980), 393-408.
20. Levy, H. Does Diversification Always Pay, in E.J. Elton and M.J. Gruber, eds., Portfolio Theory: Twenty Five Years After. Amsterdam: North Holland, 1979.

CHAPTER VII

Efficiency Measurement in Nonmarket Systems Through Data Envelopment Analysis

1. Introduction

Measurement of total factor productivity in nonmarket systems has attracted some attention in current research. Given the observed values of inputs and outputs for such systems, the problem of measurement arises due to three major reasons. First, the inputs and outputs may not have for the most part any observable market prices; hence the imputation of an equilibrium market model is not directly available. Second, the production set to which the input and output vectors belong is not directly identifiable from the observed data set; also the input-output data may be subject to a stochastic generating mechanism. Third, the units may not be profit oriented, where profit is a scalar measure of performance.

Data envelopment analysis (DEA) is a powerful technique developed by Charnes, Cooper and his associates [1-3] for measuring the relative efficiency of a set of decision-making units (DMUs), although some of its basic concepts were pioneered by Farrell and Fieldhouse [4]. This technique employs the basic notion of Pareto efficiency but in a novel manner by stipulating that a given DMU is not relatively efficient in producing its outputs from given amounts of inputs, if it can be shown that some other DMU or, combination of DMUs can produce more of some outputs without producing less of any other output and without utilizing more of any input. This technique is most suitable for application to public sector units like public schools, county hospitals and public service agencies where multiple goals cannot usually be reduced to a common single goal such as profits. Any complex system comprising different DMUs, each using a set of inputs (resources) to realize a set of outputs (objectives) may be evaluated in terms of the DEA model and the number of units which are relatively inefficient can be identified. Once the identification is done, it is possible to discuss the various alternative policies which may be undertaken to improve the relative inefficiency underlying the overall system.

The usefulness of the DEA model may be better appreciated in terms of two other methods of efficiency measurement currently available in modern economic theory. One is the economic theory of production function which expresses various outputs as specific functions of the several inputs and inefficiency is said to arise whenever given inputs lead to outputs that are less than the maximum attainable. Producion frontier is differentiated from the average production function in that it includes only the efficient units; the average production function estimated by regression methods is by implication an average of both efficient and inefficient units. The empirical construction of a multi-input multi-output production function is complicated due to problems of both specification and estimation. On the specification side, specific functional forms which are suitable for estimation require to be chosen and since the detailed data of allocation of each input to several outputs is usually not available, the frontier function is not very determinate. On the estimation side the frontier function implies that the errors when additive must lie on one side of the frontier. This leads to estimation under constrained and hence nonsymmetric errors. The standard maximum likelihood method or, the assumption of normality of additive errors around the frontier does not fit this situation very appropriately. The DEA model does not require any specific assumption about the functional form of the multi-input multi-output production frontier, yet develops a scalar measure of efficiency of a DMU as the maximum of a ratio of weighted outputs to weighted inputs subject to the condition that similar ratios for othe DMUs be less than or equal to unity. When suitably interpreted, the method of convex closure of the efficient points specified by the programming framework of the DEA model may be utilized to classify the heterogeneous data structure into two subsets, one more efficient than the other and then separate regressions may be computed over the two subsets. Such comparisons of the DEA method with the usual production frontier approach as estimated by the econometric methods [4,5] reveal the former to be more general and more precise in locating the relatively inefficient units and this appears to be true even in the stochastic case when the input-output relations are influenced in part by a stochastic generating mechanism.

The second approach to efficiency measurement, formalized in details by Debreu [6] and recently extended by Diewert [7] arises in a general equilibrium framework of the whole economy, where inefficiency may be related to the inoptimal allocation of resources. Three sources of waste in the allocation of resources are distinguished here e.g., (i) waste due to the underemployment of available physical resources, (ii) waste due to technical inefficiency in production e.g., some production units do not produce up to the maximal output level feasible by the given technology, and (iii) waste due to the imperfections of economic organization e.g., different production units while technically efficient face different prices for the same inputs or, outputs and this causes net outputs aggregated across production units to fall below what is attainable if the economic organization were efficient. The coefficient of resource utilization defined by Debreu depends however on our ability to estimate the convex hull of a generalized convex production set and interpret the optimal shadow prices in terms of a competitive framework of a general equilibrium model. Without requiring a general equilibrium framework, Hanoch and Rothschild [8] developed a similar concept of resource utilization in a standard production model and specified nonparametric and nonstatistical tests to see if the observed input-output data are consistent with a production frontier. The tests are nonparametric in the sense that they do not require the assumption of any specific function form of the input-output relation and they are nonstatistical because no assumptions about the probability distribution of the errors of observation (or disturbance terms) are introduced. Diewert and Parkan [4] have recently extended this nonparametric approach in terms of linear programming tests of consistency, where the tests are nonparametric and nonstatistical but can be applied in three situations e.g., (a) when only the data on input vectors for a single output are available for different units, (b) when both input and output vectors can be used, and (c) when the price data are available for defining suitable profit measures.

The DEA is more specific in the sense that it is not tied to a competitive equilibrium setup; it does not require that the prices of inputs and outputs are observed. The DEA model defines a convex closure of suitable input-output points in terms of piecewise linear functions, by joining which one could construct a production frontier (i.e. a convex isoquant). Thus the data requirements of this method are more modest. Two basic requirements are as follows. For each DMU the data on inputs utilized an outputs achieved must be observed and if such observations are all positive, the model necessarily defines optimal solutions i.e., either all data points are on the optimal frontier or, at least one vector point is not on the optimal frontier. The latter is not efficient relative to other DMUs. Note that in the stochastic case this method of efficiency measurement can be used as a device for stratification in to two or more strata in an efficiency scale and then separate regressions can be run over each stratum. Such attempts at combining regression and DEA approaches [3,5,10] have generally produced much better results than the straight regression model. A second requirement for the DEA model is that the overall system must comprise similar DMUs, similarity being viewed in terms of the comparability of input-output data for different DMUs. For instance one elementary school may be usefully compared to other similar elementary schools in the county, but not so usefully perhaps to the high schools. This is because the elementary school system and the high school system may not be comparable in all aspects of the input-output specification. Charnes and Cooper [2] have used the term 'managerial efficiency' while comparing DMUs in a given cluster and 'program efficiency' while comparing two different clusters. Note that the efficiency measurement by the DEA method is nonparametric, although it is not necessarily non-statistical. In suitable cases even increasing returns to scale may be handled by this method by defining suitable transformations which make the resulting production possibility set convex.

Our objective here is two-fold: to present a brief technical overview of the DEA model and its linkages with the other two models of efficiency measurement and then to consider several extensions of the DEA method for the stochastic and dynamic cases. Stochastic cases arise through a natural generalization of the deterministic production frontier into stochastic frontiers and dynamic cases are due to durable inputs like capital which may increase output only after a time lag.

2. Nonparametric Methods of Efficiency Measurement

The concept of efficiency and its measurement varies from one system to another. The theory of efficiency for economic units, e.g. a multi-product firm, is closely related to the existence of an input-output relation $f(y,x)$ between the output (y) and input (x) vectors. For the case of a scalar output and m inputs the production function $\phi(x)$ may be viewed in two ways: either as maximal output obtainable from an input vector x i.e. $\phi(x) = \text{Max } \{y | x \varepsilon L(y)\}$, where $L(y)$ denotes all input vectors yielding at least the output rate y, or alternately as the input correspondence $L(y) = \{x | \phi(x) \geq y\}$. The production function

$\phi(x)$ is usually assumed to have some regularity conditions e.g., (i) $\phi(x)$ is a nonnegative real valued function defined over the nonnegative orthant, (ii) continuous from above, (iii) nondecreasing and (iv) quasiconcave. Part (i) means that inputs and output are nonnegative satisfying perhaps the natural condition $\phi(0)=0$ and $\phi(x)>0$ for some $x>0$. Part (ii) means that the level set $L(y)$ of input correspondence is a closed set for each nonnegative y. Part (iii) excludes the possibility of decreased outputs when greater amounts of inputs are used and part (iv) means that each level set $L(y)$ is a convex set. On the basis of a production function one may easily define an efficient subset $L^*(y)$ of the input correspondence $L(y)$ as follows:

$$L^*(y) = \{x | x \varepsilon L(y), u \leq x \text{ implies that } u \notin L(y) \text{ for } y>0\} \quad (1.1)$$
$$\text{and } L^*(0) = 0$$

This says that the efficiency subset does not contain any input vector lower than x (in a least one component) yielding the output rate y. In this sense it includes the minimal input vector for a given output rate.

In the more general case when we have a production possibility set $F(x,y)$ defined by the input (x) and output vectors (y), we may have similar characterization of an efficient subset. Assume that the production set $F(x,y)$ is closed, bounded and convex for any $x \varepsilon X$, $y \varepsilon Y$ where X and Y are suitable subsets of the nonnegative orthant. Define a set $F^*(x,y^*)$ to denote the minimal inputs required to achieve at least the output vector y^*, where $F^*(x,y^*)$ is a subset of $F(x,y)$ and not empty. Let y be any other output vector not belonging to F^*, where $F^* = F^*(x,y^*)$. The distance from y to the set F^* may thus provide a measure of inefficiency, which was characterized by Debreu [6]. By definition the output vector y^* is efficient if there exists no other vector y in F* such that y≥y* with at least one component strictly greater. Since the production set F* is convex, and y* is an efficient point in F*, therefore there must be a vector p of nonnegative prices p≥0 such that

$$p \cdot (y^* - y) \geq 0 \text{ for all } y \; \varepsilon \; Y \tag{1.2}$$

Denote by y^0 a vector collinear with y but belonging to the set F* i.e. $y^0 = ry$, then it follows:

$$\max_{y^* \varepsilon F^*} \frac{p \cdot y}{p \cdot y^*} = \frac{1}{r} \max_{y^* \varepsilon F^*} \frac{p \cdot y^0}{p \cdot y^*} \leq \rho \tag{1.3}$$

where $\rho = 1/r$ is called the coefficient of resource utilization and the ratio $(p \cdot y^0 / p \cdot y^*)$ equals unity when $y^0 = y^*$. The coefficient ρ of resource utilization then attains its maximal value 1.0; In all other cases of $\rho < 1 \cdot 0$ we have inefficient resource utilization in the input-output system, resulting in a "deadweight loss". Two limitations of the above concept of a nonnegative price vector p≥0 associated with y in (1.2) have been noted by Debreu. One is that in many cases no price vector may exist at all in some concrete economic situations e.g., there may be no uniqueness of price for each commodity; second, the price vector may have no intrinsic economic significance e.g. it may not be readily used as market prices to compute profit. Note also that the minimum feasibility set F* may not be very easy to calculate in any actual situations, unless it can be suitably expressed as computable mathematical programs. Diewert [7] and Diewert and Parkan [9] have developed suitable characterizations of this concept of efficiency in terms of linear programming (LP) models. Thus consider the single output case with m inputs for each DMU. Given n observations (x_j, y_j) for j = 1,2,...,n we ask the following question: does there exist a production function $y = f(x)$ with some regularity conditions, such that y_j is the maximal amount of output that can be produced from the vector of inputs x_j for $j=1,2\ldots,n$? Suppose the regularity conditions on $f(x)$ are those discussed in connection with (1.1) and assume that the observed output data are distinct i.e., $0 \leq y_1 < y_2 < \ldots < y_n$. Then Diewert and Parkan [9] consider the following LP problem in the following nonnegative variables (μ_i, θ_{i+1}, θ_{i+2}, ..., θ_n) denoted by ψ:

$$\min_{\psi \geq 0} \{\mu_i : \sum_{j=i+1}^{n} \theta_j x_j \leq \mu_i x_i, \sum_{j=i+1}^{n} \theta_j = 1\} \tag{1.4}$$

where $i=1,2,\ldots,n-1$. It is clear that if the m-element vector x_i is strictly positive, then an optimal feasible soluation to the LP model (1.4) must exist and the optimal objective function denoted by μ_i^* is bounded from below by zero. Hence they proposed the following consistency test:

Test of Consistency

If $\mu_i^* > 1$ for $i=1,2,\ldots,n-1$ where μ_i^* is optimal in the LP problem (1.4), then the observed data set $\{x_j, y_j; j=1,2,\ldots,n\}$ is consistent with the efficiency hypothesis that there exists a production function $f(x)$ such that y_j is the maximal amount of output obtainable from the vector of m inputs x_j for each $j=1,2,\ldots,n$.

It follows by implication that if $\mu_i^* < 1$ for any $i=1,2,\ldots,n-1$ then the efficiency hypothesis fails for the given data set. The test of consistency with the efficiency hypothesis can be alternatively stated in terms of the dual of the LP models (1.4) as follows:

$$\text{Max}_{\alpha_i, w_i \geq 0} \{\alpha_i : w_i \cdot x_i \leq 1, \ \alpha_i \leq w_i \cdot x_j, \ j=i+1,\ldots,n\} \quad (1.5)$$

$$\text{where } w_i \cdot x_i \equiv \sum_{k=1}^{m} w_{ik} x_{ik}$$

Let α_i^* be the finite maximal value of this dual LP model. Then by the strong duality theorem of LP we must have $\mu_i^* = \alpha_i^*$ for $i=1,2,\ldots,n-1$.

If $\mu_i^* < 1$ for any i, then it can be interpreted as Debreu's coefficient of resource utilization and $\delta_i^* = 1 - \mu_i^*$ as Debreu's measure of economic loss due to the inefficient utilization of resources or inputs. Two implications of the test of consistency with the efficiency hypohtesis must be noted. One is that the actual outputs y_j do not appear in any of the two LP models. This means that the consistency test do not require any detailed knowledge about the magnitude of the y_j in each period: all that is required is an ordinal ordering of outputs as $0 < y_1 < y_2 < \ldots < y_n$ and the corresponding labels for the input vectors e.g., x_k corresponds to y_k where $y_{k-1} < y_k < y_{K+1}$ for $k=1,2,\ldots,n-1$. The second implication of the consistency test is that it suggests a linear transformation of the data set in case the efficiency hypothesis fails, so that the transformed data set can satisfy the efficiency hypothesis. Thus, in case $0 < \mu_i^* < 1$ for some i we replace the input vector x_i by the shrunken input bector $(\mu_i^* - \varepsilon) x_i$ for any ε such that $0 < \varepsilon < \mu_i^*$, then the optimal solution of the transformed programming problem would satisfy the efficiency hypothesis for observation i. Hence if $\mu_i^* < 1$, then $\delta_i = 1 - \mu_i^*$ is a natural measure of violation of the efficiency hypothesis for the i-th input vector x_i. Such measures of inefficiency were proposed earlier by Hanoch and Rothschild [8] and by Afriat [11]. Afriat's contribution noted specifically the importance of the stochastic case by acknowledging that actual production operations may not be perfectly efficient in all situations and hence the production function which represents the maximum output has to be inferred from some probabilistic hypothesis about the distribution of efficiency. Let $p(e)$ be the probability density of efficiency $e=y/y^*$ of an operation (x,y), where it is assumed to be independent of the input levels x. Let $y_1, y_2,\ldots,y_n$ be the feasible output levels and y^* the maximum possible output. Then we have to determine the estimate $\hat{y}^*$ of maximum output so as to make the likelihood $L(y_1,y_2,\ldots,y_n|y^*) = \prod_{t=1}^{n} p(y_t/y^*)$ a maximum. Because of the nonnegativity restrictions on inputs and outputs, the beta distribution is considered to be a plausible model for the probability distribution of the efficiency ratios $e = y/y^*$.

Now consider the DEA model, where each of n DMUs has input output vectors (x_j, y_j) of dimensions m and r respectively and we define the efficiency ratio by h_j:

$$h_j = (\sum_{i=1}^{r} u_i y_{ij}) / (\sum_{s=1}^{m} v_s x_{sj}) \qquad (2.1)$$

$s=1,2,\ldots,m; \; i=1,2\ldots,r; \; j=1,2,\ldots,n$

Let k be the reference DMU which is to be compared in efficiency relative to other DMUs in the cluster. Then we solve for the optimal values of vectors u and v:

$$\max_{u,v\geq 0} \{h_k = h_k(u,v): h_j \leq 1; \; u,v \geq 0; \; j=1,2,\ldots,n\} \qquad (2.2)$$

For any fixed k, $k-1,2,\ldots,n$ this involves solving mathematical programs which are known as linear functional fractional programs, which by suitable transformations can be reduced to LP models [1-2]. Let $u^*=u^*(k)$ and $v^*=v^*(k)$ be the optimal solution vectors for the k-th DMU belonging to the cluster of n units and h_k^* be the maximal value of the objective function. It can be shown that if the observed input and output vectors are all positive and known constants i.e. they imply optimal behavior in the given environment, then the optimal set $\{u^*(k), v^*(k), h_k^*\}$ would exist.

If for any k, $1 \leq k \leq n$ we have $0 < h_k^* < 1$, then the k-th DMU is not efficient relative to others in the cluster and hence the entire set of input-output data is not consistent with the efficiency hypothesis. Thus $\delta_k^* = 1 - h_k^*$ may be used as a measure of inefficiency i.e. a measure of violation of the efficiency hypothesis for the k-th unit. Two implications of this measure are useful in economic terms. One is that h_k^* is associated with the optimal vectors $u^*(k)$, $v^*(k)$ which may be interpreted either as optimal weights or optimal shadow prices. Thus if h_k^* cannot attain the value 1.0, which characterizes the efficient DMUs, it cannot be efficient for any other optimal vectors $u^*(q)$, $v^*(q)$ where $q \neq k$ and $1 \leq q \leq n$. One could thus use this efficiency measure to rank the set of n DMUs in an efficiency scale from the lowest to the highest, provided the reference DMU is representative. Secondly, the DEA model (2.2) clearly defines a nonparametric procedure through its specification of the convex closure of the efficient points. This can then be directly compared with the production frontier approach or the statistical approach of an average production function.

Suppose the representative DMU has been selected from the set of efficient units in the cluster with input-output vectors $(\bar{x}, \bar{y})$, then the n DMUs can be ranked by the programming model:

$$\underset{\bar{u}, \bar{v} \geq 0}{\text{Max}} \ \{\bar{h} = h(\bar{u}, \bar{y}): h_j \leq 1, \ j=1,2,\ldots,n\} \qquad (2.3)$$

where $h_j = h_j(\bar{u}, \bar{y}) = (\sum_{i=1}^{r} \bar{u}_i \bar{y}_i) / (\sum_{s=1}^{m} \bar{v}_s \bar{x}_s)$. The set $(\bar{u}^*, \bar{v}^*)$ of optimal vectors may then be utilized in several ways. First of all, if $\bar{u}_i^*$ and $\bar{v}_s^*$ are positive for each i and s, then one can interpret these as suitable prices of outputs and

inputs and a pseudo-profit function for the representative DMU: $\bar{\Pi}^* = \sum_{i=1}^{r} \bar{u}_i^* \bar{y}_i - \sum_{s=1}^{m} \bar{v}_s^* \bar{x}_s$ may be imputed, where $\frac{\partial \bar{\Pi}^*}{\partial x_s} = \sum_{i=1}^{r} \bar{u}_i^* \frac{\partial \bar{y}_i}{\partial \bar{x}_s} - \bar{v}_s^* = 0$
defines the usual pricing rule of competitive markets. So long as the addition of new units or deletion of existing units does not alter the mean input output vectors $(\bar{x}, \bar{y})$ of the representative DMU, the same objective function $\bar{h} = h(\bar{u}, \bar{v})$ may be used to rank the transformed cluster in an efficiency scale. Thus one may define a subset S_0 of the efficient set S to be robust, if additions of new units or, deletions of existing ones do not alter the maximum efficiency character of the units included in S_0. It is clear that such a robustness property is unlikely to be realized for arbitrary additions or deletions (truncations). Let D be the domain of data variations $(X,Y) \varepsilon D$ for which the subset S_0 retains 100% efficiency. The number of DMUs belonging to this subset S_0 may then be called D-robust. By including the D-robust efficient units alone in our sample, one could run linear regressions of the weighted outputs $y_j^* = \sum_{i=1}^{r} u_i^* y_{ij}$ on the inputs $x_j = (x_{1j}, x_{2j}, \ldots, x_{mj})$ and the R^2 value would be very close to one. As more and more efficient units which are not D-robust are included in our regression sample, the R^2 value would decline. Furthermore if inefficient units are also included, the R^2 value may decline still further. Thus the DEA model provides a rationale for stratifying the n sample DMUs into suitable subsets, whereby the frontier production function may be estimated by regression with a high R^2 value.

3. Generalizations of the DEA Model

Two aspects of the DEA model method not yet fully explored are robustness issues under stochastic variations of observed data and the presence of capital inputs with gestation lags of one or more time periods. Thus, a particular DMU which may be relatively inefficient (efficient) may indeed turn otherwise if the input output data are filtered of their noisy components. Similarly, if some of the m inputs are used for capital accumulation or, investment this period, its impact on increased output may be estimated only in later periods. Hence additional variables denoting incremental outputs and inputs must be incorporated, along with depreciation of capital inputs which may differ across different inputs.

For simplicity we consider here the framework where each DMU has one output but m inputs and the DEA method seeks to determine a vector of optimal weights in terms of which any reference unit can be compared in efficiency.

Let x_{ij} denote the quantity of input $i=1,2,..,m$ and y_j be the single output for j-th DMU $(j=1,2,...,n)$. Then the DEA model may be simply represented by the following LP model:

$$\min_{\beta} g_k = \sum_{i=1}^{m} \beta_i x_{ik}$$

subject to (3.1)

$$\sum_{i=1}^{m} \beta_i x_{ij} \geq y_j; \quad j=1,2,\ldots,n$$

$$\beta_i \geq 0; \quad i=1,2,\ldots,m$$

By letting the reference DMU index k vary over $1,2,..,n$ one may determine n optimal weights $\beta^*(k)$, $k=1,2,..,n$ for the inputs. For the more general case of s outputs $(s \geq 1)$, the DEA model becomes

$$\min_{\beta,\alpha} g_k = \sum_{i=1}^{m} \beta_i x_{ik}$$

subject to (3.2)

$$\sum_{i=1}^{m} \beta_i x_{ij} \geq \sum_{r=1}^{s} \alpha_r y_{rj}; \quad j=1,2,\ldots n$$

$$\sum_{r=1}^{s} \alpha_r y_{rk} = 1$$

$$\beta_i \geq 0, \ i=1,2,\ldots,m; \ \alpha_r \geq 0, \ r=1,2,\ldots,s$$

This model is derived by Charnes, and Cooper [1,2] from a more general formulation in terms of linear functional fractional programming models and they have shown the equivalence of this LP model with other formulations due to Farrell and Fieldhouse [4], and others [11].

In empirical applications the observed input and output quantities denoted by $X = (x_{ij})$, $y = (y_j)$ are generally positive for all i and j. In such cases the LP model (1) is always feasible, as it is obvious from the dual:

$$\max_{\lambda} z = \sum_{j=1}^{n} y_j \lambda_j$$

subject to (3.3)

$$\sum_{j=1}^{n} x_{ij} \lambda_j \leq x_{ik}; \quad i=1,2,\ldots,m$$

$$\lambda_j \geq 0; \ j=1,2,\ldots,n$$

Hence there must exist an optimal feasible vector $\beta^* = \beta^*(k)$ for any given k.

The primal and dual problems may be written in vector-matrix terms as

$$\min_{\beta} g_k = X_k'\beta$$

$$\text{subject to } X'\beta \geq y \qquad (3.4)$$

$$\beta \geq 0$$

and

$$\max_{\lambda} z = z_k = y'\lambda$$

$$\text{subject to} \qquad (3.5)$$

$$X\lambda \leq X_k$$

$$\lambda \geq 0$$

Here prime denotes transpose, X_k is the m-element column vector (x_{ik}), $X = (x_{ij})$ the m by n input-output matrix and $y = (y_j)$ is the n-element column vector. The decision variables are the column vectors β and λ with m and n elements respectively. Two types of interpretations have been given for the decision vectors such as β. One is that they represent a set of nonnegative weights which facilitate the efficiency ranking and comparison of DMUs in the cluster of n units. Recently Nunamaker [15] has considered some specific types of data variations (e.g., additional data which are perfectly positively correlated with one or more existing model variables) as they affect the weighting scheme. A second approach is to interpret the β^* coefficients as the parameters of a suitable production function, which can then be easily related to the econometric theory of the production frontier developed by Timmer [13] and Aigner and Chu [14]. Recently an attempt has been made by Banker, Charnes, Cooper and Schinnar [3] to incorporate the problem of statistical estimation of a production frontier in the DEA model. The data variation problem they consider arises in those situations where one may have reason to assume that the concept of a parametric production function (e.g., output is a linear function of the inputs with all variables measured in logarithmic units) is reasonable for the n DMUs and the object is to estimate the parameters of the frontier production function by a DEA principle.

3.1 Stochastic Aspects

We consider first data variations which affect the coefficients of the objective function only. For the primal problem (3.4), this means that the vector X_k changes for different k, k=1,2,..,n and this may result in different sets of values of the vector β^*. Now if the data variations in X_k are due to stochastic factors, they can be analyzed in terms of their stochastic generating mechanisms. For instance if the DMUs refer to a homogeneous cluster of n schools, the input data would tend to center around a mean or median level with a certain degree of dispersion around the central location. The data generating model in this case may take the following form:

$$X_k = \bar{x} + \varepsilon_k \qquad (3.6)$$

where $\bar{x} = (\bar{x}_i)$ is the mean (or median) input vector (i.e. for the mean $\bar{x}_i = (1/n) \sum_{j=1}^{n} x_{ij}$) and ε_k is a random vector which may follow any of several multivariate probability distributions. If the probability distribution of ε_k is such that it has mean zero and a finite variance-covariance matrix V_ε, then one can easily characterize how closely the input vector points X_k are scattered around its mean $\bar{x}$. This is done through the multivariate distance function $D^2 = D^2(\tilde{x}; \bar{x}, V_\varepsilon)$ defined as follows:

$$D^2 = D^2(\tilde{x};\ \bar{x}, V_\varepsilon) = (\tilde{x}-\bar{x})V_\varepsilon^{-1}(\tilde{x}-\bar{x}) \tag{3.7}$$

where it is assumed that the variance-covariance matrix V_ε is nonsingular and the notation $\tilde{x}$ has been used in place of X_k. Thus if all X_k in the cluster are very close to the mean, the value of D^2 will be close to zero. The higher the value of D^2, the farther the points are scattered away from the mean level. The role of the variance-covariance matrix V_ε is more subtle however. It has two practical implications. One is that it acts as a filtering device, whereby with noisy data any difference $\tilde{x}_i - \bar{x}_i$ is deflated or corrected in terms of its standard errors, so that the squared distance between $\tilde{x}$ and $\bar{x}$ can be more reliably computed. Secondly, if the underlying probability distribution of the errors ε can be assumed to be normal (e.g. under certain conditions by the central limit theorem), then the random quantity D^2 in (3.7) is known to follow a chi-square distribution with m degrees of freedom if the variance-covariance matrix V_ε is known. Thus one may choose a level of significance α and determine a suitable positive scalar $c=c_\alpha$ such that $\tilde{x}$ defines a confidence region

$$\tilde{X} = \{\tilde{x} | D^2(\tilde{x};\ \bar{x},\ V_\varepsilon) \leq c_\alpha\} \tag{3.8}$$

in the sense that for any set of points x belonging to region $\tilde{X}$ we have $\text{Prob}(x\varepsilon\tilde{X}) = \alpha$. By varying α one can therefore vary the size of the neighborhood around $\bar{x}$. Now we replace the vector X_k in the objective function of the efficiency model (3.4) by the mean vector $\bar{x}$ defined in (3.6) and set up the deterministic LP model

$$\min_{\beta \in R} \bar{g} = \bar{x}'\beta \qquad (4.1)$$

where $R = \{\beta | X'\beta \geq y; \beta \geq 0\}$

This formulation may be very useful in several ways. First, it leads to only one set of optimal weights (assuming unique optimality). Thus, if the n DMUs are individual schools, e.g., the objective function represents the weighting function of the overall school system administrator. This represents the industry manager's efficiency problem comprising n units or enterprises. Second, one may choose points $\tilde{x}$ which are close to $\bar{x}$ in the sense defined by (3.8), thus generating new solutions which are either the same as before or different. In the first case we have robust solutions within the neighborhood defined by (3.8). Third, one may write the dual problem of (4.1) as

$$\max_{\lambda} \bar{z} = y'\lambda$$

subject to $\qquad (4.2)$

$$X\lambda \leq \bar{x}; \quad \lambda \geq 0$$

Maximizing the value of output $\bar{z}$ for the industry or the cluster under the constraint of average input availability shows very clearly the industry manager's efficiency problem i.e. he has to select the efficient units in such a way that the output value $\bar{z}$ is maximized. Suppose we consider points $\tilde{x}$ very close to $\bar{x}$ such that the optimal basis matrix remains the same for the solution vector $\bar{\beta}^*$ of the LP model (4.2). Then it is very easy to show that

$$E\bar{z}^*(\tilde{x}) \leq \bar{z}^*(\bar{x}) \qquad (4.3)$$

where the expectation operator E is over the probability distribution of vector $\tilde{x}$ with its expectation $E(\tilde{x}) = \bar{x}$. The inequality (4.3) is strict i.e. $E\tilde{z}^*(\tilde{x}) < \bar{z}^*(\bar{x})$, where asterisk denotes maximal values, provided the probability distribution of vector $\tilde{x}$ is nondegenerate. This has implications for the industry efficiency problem e.g., the industry manager may find it more efficient to choose a mean input level $\bar{x}$ of a representative unit.

The sensitivity of the optimal objective function $\bar{g}^*(\bar{x})$ of the LP problems (4.1) may also be directly analyzed in terms of data variations affecting the mean input vector $\bar{x}$. Since the minimand $\bar{g}^*(\bar{x})$ is a continuous function of $\bar{x}$ at the optimal solution vector $\bar{\beta}^*$, there must exist a neighborhood $N(\tilde{x}; \bar{x})$ of $\bar{x}$:

$$N(\tilde{x}; \bar{x}) = \{\tilde{x} | \bar{x}_a \leq \tilde{x} \leq \bar{x}_b\} \qquad (4.4)$$

where $\bar{x}_a$, $\bar{x}_b$ are two suitable fixed points such that for any point $\tilde{x}$ belonging to $N(\tilde{x}; \bar{x})$, the original solution vector $\bar{\beta}^*$ remains optimal. Now let A be the intersection of the two sets $\tilde{X}$ and $N(\tilde{x}; \bar{x})$, where $\bar{x}_a$, $\bar{x}_b$ are fixed and α is varied in (3.8) such that A is a nonempty set. It is easy to show that such a construction is always feasible for a given solution vector $\bar{\beta}^*$. For instance we may write (4.3) as the component-by-component restrictions:

$$\bar{x}_{ai} \leq \tilde{x}_i \leq \bar{x}_{bi}, \quad i=1,2,\ldots,m$$

which define an m-dimensional cuboid as follows

$$\frac{|\tilde{x}_i - (\bar{x}_{ai} + \bar{x}_{bi})/2|}{\tfrac{1}{2}(\bar{x}_{bi} - \bar{x}_{ai})} \leq 1, \quad i=1,2,\ldots,m \qquad (4.5)$$

340   CHAPTER VII

Now we construct an ellipsoid $(\tilde{x}-\tilde{x}_0)'T(\tilde{x}-\tilde{x}_0) = 1$ which contains the cuboid and fulfills the following conditions

(i) the ellipsoid has the same center

$$x_0' = (\frac{\bar{x}_{a1}+\bar{x}_{b1}}{2}, \frac{\bar{x}_{a2}+\bar{x}_{b2}}{2}, \ldots, \frac{\bar{x}_{am}+\bar{x}_{bm}}{2})$$

as the cuboid

(ii) the axes of the ellipsoid are parallel to the coordinate axes i.e. the matrix T of the ellipsoid is diagonal i.e. T = diag. $(t_1, \ldots, t_m)$

(iii) the surface of the ellipsoid contains all corner points of the cuboid, and

(iv) the ellipsoid has the minimal volume

$$V = \pi \prod_{i=1}^{m} (1/t_i)$$

This leads to the optimizing problem

$$\min_{\{t_i\}} V \text{ subject to } \sum_{i=1}^{m} (\frac{\bar{x}_{ai}-\bar{x}_{bi}}{2})^2 t_i = 1$$

the optimal solution for which gives

$$t_i = (4/m)(\bar{x}_{ai}-\bar{x}_{bi})^{-2}, \quad i=1,2,\ldots,m$$

$$x_0' = (\frac{\bar{x}_{a1}+\bar{x}_{b1}}{2}, \frac{\bar{x}_{a2}+\bar{x}_{b2}}{2}, \ldots, \frac{\bar{x}_{am}+\bar{x}_{bm}}{2})$$

$$T = \text{diag}(t_1, t_2, \ldots, t_m)$$

The cuboid (4.4) is thus contained in the ellipsoid: $\tilde{X}_0 = \{x | (\tilde{x}-\tilde{x}_0)' T(\tilde{x}-\tilde{x}_0) \leq 1\}$. Now if we let the center $\bar{x}$ of the ellipsoid (3.8) equal the center $\tilde{x}_0$ of the cuboid (4.4) we obtain

EFFICIENCY MEASUREMENT IN NONMARKET SYSTEMS

$$\tilde{X}_0 = \{x \mid (\tilde{x}-\bar{x})'T(\tilde{x}-\bar{x}) \leq 1\}$$

$$\tilde{X} = \{x \mid (\tilde{x}-\bar{x})'V_\varepsilon^{-1}(\tilde{x}-\bar{x}) \leq c_\alpha\} \quad (4.5)$$

since c_α can vary in the positive domain $0 < c_\alpha < \infty$, the intersection $A = \tilde{X} \cap \tilde{X}_0$ of the two sets cannot be empty. Hence we can state the following result.

Theorem 1

There always exists an optimal solution vector $\bar{\beta}^*$ of the industrial efficiency model (4.1), if the inputs (x_{ij}) and outputs (y_j) are positive. And if the data variations $(\tilde{x})$ around the mean input level $\bar{x}$ are such that $\tilde{x} \varepsilon A$, where A is the intersection of the two sets $\tilde{X}_0$ and $\tilde{X}$ defined in (4.5), then the solution vector $\bar{\beta}^*$ is robust.

Two implications of this result may be noted. First, the ellipsoid $\tilde{X}$ may be viewed as a normal theory approximation of $\tilde{X}_0$, provided the data variations in ε are suitably trimmed so that V_ε^{-1} approximates T. In this sense the probability $\tilde{x} \varepsilon A$ may be directly computed and significance tests performed. Second, for any outlying observation i.e. $\tilde{x} \varepsilon A$ one may easily compute a point $\tilde{x} \varepsilon A$ which has the shortest distance in the sense of minimum $D^2(\tilde{x}, \bar{x}, V_\varepsilon)$.

Next we consider data variations in output which affect the objective function of the dual problem (4.2). Such data variations may arise in practical situations in two ways. One arises when the output measures have large and uncertain measurement errors, which are much more significant than the input measures. For example in school efficiency studies, the input costs like teacher salaries, administrative expenses etc. may have low measurement errors, whereas the performance test scores of students may contain large errors of measurement of true student quality. Second, many of the outputs may be internal to a DMU, reflecting noncooperative or rival behavior among its constitutents.

Data variations in output may be represented in at least two ways:

$$y_j = \bar{y} + \theta_j \quad \text{(homoscedastic)} \quad (5.1)$$

$$y_j = \bar{y}_j + \theta_j \quad \text{(heteroscedastic)} \quad (5.2)$$

In the first case outputs of different units have the same mean but different variances (i.e. $E\theta_j = 0$ for all j and $E\theta_j^2 = \sigma_j^2 > 0$), whereas in the second case the mean levels $\bar{y}_j$ of output are also different. The latter situation would arise if for each DMU we have time series data on outputs, so that $\bar{y}_j$ represents the time-average of output of DMU $j = 1,2,..,n$. Since (5.1) can be subsumed as a special case of (5.2), we would consider the heteroscedastic case only when the output vector y is assumed to be distributed with a mean vector $\bar{y}$ and a variance-covariance matrix V_θ.

The industry manager's problem in maximizing the value of output in (4.2) by selecting the most efficient units is now complicated by two stochastic factors. One is that any large variability in output (measured by V_θ) would bias the method of efficiency measurement and ranking. Second, any lack of knowledge of the probability distribution of outputs would introduce a specification bias e.g. we assume the distribution to be normal, although it may not be so. We consider the first case here and introduce a concept of risk-averse efficiency due to Peleg and Yaari [16]. Thus we define the industry manager's utility function $u = u(\tilde{z})$, $\tilde{z} = \tilde{y}'\lambda$ to belong to a set $U(\cdot)$ of concave and nondecreasing utility functions and consider a risk-adjusted preference function $f = f(\bar{z},\sigma^2)$ for ordering the decision vectors in a stochastic environment. For the class of concave utility functions with a constant rate of absolute risk aversion, this leads to the risk preference function

$$f = \bar{z} - \frac{r}{2}\sigma^2 \ ; \ r \geq 0$$

where $\bar{z} = \bar{y}'\lambda$, $\sigma^2 = \lambda'V_\theta\lambda$ and r is the Arrow-Pratt measure of absolute risk aversion. The risk preference function tells us how much of his expected return the industry manager faced with a risky prospect would be willing to sacrifice to an insurance company in order to achieve certainty. The final decision problem corresponding to the LP model (4.2) now becomes

$$\max_{\lambda \varepsilon C} f = \bar{y}'\lambda - \frac{r}{2}\lambda'V_\theta\lambda \qquad (5.3)$$

where $C = \{\lambda | X\lambda \leq \bar{x}; \ \lambda \geq 0\}$. For any nonnegative r this quadratic programming (QP) model has an optimal solution $\lambda^* = \lambda^*(r)$. The dual QP problem is in terms of two sets of decision variables $\lambda(r)$ and $\beta = \beta(r)$:

$$\min_{\lambda,\beta} g(r) = \bar{x}'\beta + (r/2)\lambda'V_\theta\lambda$$

s.t. (5.4)

$$X'\beta \geq \bar{y} - rV_\theta\lambda$$

$$\beta \geq 0, \ \lambda \geq 0$$

Two aspects of the QP models (5.3), (5.4) are to be noted when we compare them with their LP counterparts (i.e. r=0). First, the optimal efficiency ranking is now in terms of two sets of variable $\beta^*(r)$ and $\lambda^*(r)$ with r>0 rather than one and since the objective function is quadratic, the optimal solutions $\beta^*(r)$ for example need not be an extreme point of the set $\{X'\beta \geq \bar{y}, \beta \geq 0\}$ i.e. unlike the LP model, more than m elements (m<n) of $\beta^*(r)$ may be positive. Second, the optimal quadratic solution $\beta^*(r)$, $\lambda^*(r)$ would be more stable than the corresponding LP solution, when stability is viewed in terms of variance of return $\sigma^2 = \lambda' V_\theta \lambda$. The reason for this is that the optimal QP solutions are more diversified in terms of risk than the optimal LP solutions.

We have thus arrived at two generalizations of the DEA model of linear efficiency. One specifies through a pair of QP models a measure of quadratic efficiency which allows more diversificaiton in the optimal weights selected. The second utilizes the concept of risk-averse efficiency in the sense that by increasing r the manager aims to achieve a certainty-level of return with a high probability. By contrast, decreasing the risk-aversion parameter r to zero, he increases the cost of risk.

Next we consider the data variations in the constraints and in the general case of constraint variations, we have both inputs x_{ij} and outputs y_j subject to stochastic generating mechanisms. Hence the constraints now appear as

$$\sum_{i=1}^{m} \beta_i \tilde{x}_{ij} \geq \tilde{y}_j; \beta_i \geq 0 \qquad (6.1)$$

where $\tilde{x}_{ij} = \bar{x}_i + \varepsilon_{ij}$, $\tilde{y}_j = \bar{y}_j + \theta_j$ and the zero-mean stochastic components ε_{ij}, θ_j are such that the observed data $\tilde{x}_{ij}$, $\tilde{y}_j$ on inputs and outputs are all positive. This implies that a feasible vectove β of weights always exists. But since the constraint $\tilde{X}'\beta \geq \tilde{y}$ is now probabilistic, there is a finite probability of violating the constraints for some $j=1,2,\ldots,n$. By the theory of chance-constrained LP models developed by Charnes and

Cooper and others [17], each DMU may preassign a specified probability u_j, $0 < u_j < 1$ that the constraint j is not violated. This means

$$\text{Prob}[\sum_{i=1}^{m} \beta_i \tilde{x}_{ij} \geq \tilde{y}_j] = u_j$$

$$j = 1, 2, \ldots n$$

(6.2)

where u_j may be interpreted as a measure of reliability or tolerance for the j-th constraint. The probability of violation (or unreliability) of the j-th constraint would then be

$$1 - u_j = \text{Prob}[\sum_{i=1}^{m} \beta_i \tilde{x}_{ij} < \tilde{y}_j]$$

(6.3)

Thus one may note two basic differences of the chance-constrained case from the certainty case where $u_j = 1.0$ for every j. First, there is the certainty case. Thus if we write

$$\sum_{i=1}^{m} \beta_i \tilde{x}_{ij} + e_j^+ - e_j^- = \tilde{y}_j$$

$$e_j^+ \geq 0, \ e_j^- \geq 0, \ j = 1, 2, \ldots, n$$

(6.4)

the deviations e_j^+ are assumed here to occur with a zero probability, whereas e_j^- occurs with probability one. Second, if the reliability measure u_j are not preassigned, but they can take any value $0 \leq u_j \leq 1$, then one could write (6.4) as

where
$$\tilde{y}_j = \sum_{i=1}^{m} \beta_i \tilde{x}_{ij} + \tilde{e}_j$$
$$e_j = e_j^+ - e_j^-; \quad j=1,2,\ldots,n \qquad (6.5)$$

and estimate the average production function on the basis of n observations. If the least squares (LS) method is followed one would minimize $\sum_{j=1}^{n} e_j^2$ to obtain the LS estimates $\hat{\beta}_{LS}$. An alternative method would be to minimize the expression $\sum_{j=1}^{n} |e_j|$ which is the sum of absolute deviations. Requiring $\hat{\beta}_i \geq 0$ would make the LS method one of quadratic programming, whereas minimizing the sum of absolute deviations under $\hat{\beta}_i \geq 0$ would still preserve the LP framework.

The optimal decision problem for the industry manager concerned with the industry-level efficiency problem is not to preassign the reliability levels u_j but to choose an optimal level of system reliability defined e.g. by $R_o = \prod_{j=1}^{n} u_j$. This leads to the minimax principle of determining the efficiency measures as follows

$$\min_{\beta} \max_{u} g = \bar{x}'\beta + R_0(u_1, \ldots, u_m)$$

$$\text{s.t.} \quad \text{Prob}[\sum_{i=1}^{m} \beta_i \tilde{x}_{ij} \geq \tilde{y}_j] = u_j \qquad (6.5)$$

$$0 \leq u_j \leq 1; \quad R_0 = \prod_{j=1}^{n} u_j$$

$$\beta_j \geq 0; \quad j=1,2,\ldots,n$$

Given the observed data set $\tilde{x}_{ij}$, $\tilde{y}_j$ one could solve this problem by a repeated two-stage procedure known as reliability programming [8,9]. For instance one may start with an initial vector $u^{(1)} = (u_1^{(1)}, u_2^{(1)}, \ldots, u_n^{(1)})$ preassigned and then solve the nonlinear progarm

$$\min_{\beta} g = \bar{x}'\beta$$

$$\text{s.t.} \quad \text{Prob}\left[\sum_{i=1}^{m} \beta_i \tilde{x}_{ij} \geq \tilde{y}_j\right] = u_j^{(1)}$$

$$\beta \geq 0, \quad 0 \leq u_j^{(1)} \leq 1$$

Let $\beta^*(u^{(1)}) = \beta^*(1)$ denote the optimal solution in this stage. Given $\beta^*(u^{(1)}) = \beta^*(1)$ we next solve for a new vector u which maximizes the system reliability $R_0 = \prod_{j=1}^{n} u_j$ subject to $\text{Prob}\left[\sum_{i=1}^{m} \beta_i^*(1)\tilde{x}_{ij} \geq \tilde{y}_j\right] = u_j$; $0 \leq u_j \leq 1$. Denote this second stage solution by $u_*^{(2)} = (u_{1*}^{(2)}, u_{2*}^{(2)}, \ldots, u_{n*}^{(2)})$. We repeat this procedure till $\beta^*(t) \cong \beta^*(t+1)$ and $u_*^{(t)} \cong u_*^{(t+1)}$ for some iteration t. One can also prove the folloiwng result.

Theorem 2

If there exists an optimal solution $\beta^*(u^0)$ for a preassigned positive tolerance vector u^0, $0 \leq u_j \leq 1$, all j and the joint probability distribution of $\tilde{x}$ and $\tilde{y}$ is continuous, then a minimax solution $[\beta^{**}=\beta^*(u^*), u^*]$ exists, which solves the nonlinear problem (6.5) above.

Proof

The optimal solution vector $\beta^*(u^0)$ must satisfy

$$F_j(\beta^*(u^0)) = 1-u_j^0, \quad u_j^0 > 0$$

under the given conditions, where $F_j(\cdot)$ denotes the cumulative (marginal) density $\text{Prob}[\Sigma_i \beta_i \tilde{x}_{ij} \leq \tilde{y}_j]$. Hence in vector notation $F(\beta^*(u^0)) = \underline{1}-u^0$, where

$\underline{1}$ denotes a vector with unit elements. Since the cumulative density is continuous and increasing, it has an inverse i.e. $\beta^*(u^0) = F^{-1}(\underline{1}-u^0)$. Hence there is a compact neighborhood $N(u^0)$ around u^0 throughout which $\beta^*(u^0)$ would remain optimal. Now the system reliability measure $R_0 = R_0(u)$ is a continuous function defined on this neighborhood $N(u^0)$ and therefore it must have a maximum. Now vary u^0 within its compact domain $U = \{u|0 \geq u_j \leq 1, j=1,2,..,1\}$, till the global maximum of $R_0(u)$ is reached. Since $\log R_0 = \sum_{j=1}^{n} \log u_j$ is a strictly concave function of $u \varepsilon U$, there must exist a global maximizing vector u^* with $\beta^*(u^*) = F^{-1}(\underline{1}-u^*)$.

This result is useful in trade-off analysis between reliability improvement and cost minimization. In suitable cases even the LP structure may be maintained. For instance let $\tilde{y}_j$ be only random and not $\tilde{x}_{ij}$ and let the probability density of $\tilde{y}_j$ be a two-parameter exponential density.

$$p(\tilde{y}_j) = k_j \exp\{-k_j(\tilde{y}_j - m_j)\}, \quad \tilde{y}_j \geq m_j \geq 0 \qquad (6.6)$$

where $m_j = \min_j \tilde{y}_j$, $\tilde{y}_j > 0$ is assumed known. The minimax model (6.5) may then be written as an LP formulation:

$$\min_{\beta} \max_{v} g = \sum_{i=1}^{m} \bar{x}_i \beta_i - \sum_{j=1}^{n} v_j$$

s.t.

$$\sum_{i=1}^{m} \beta_i x_{ij} - m_j \leq v_j/k_j$$

$$v_j = -\log u_j; \quad 0 \leq u_j \leq 1, \quad j=1,2..,n$$

$$\beta_i \geq 0, \quad i=1,2,..,m$$

If the parameters k_j and m_j are estimated from the observed samples, this defines an LP model. It may be noted that Farrell and Fieldhouse [4] found in their empirical study of production efficiency of a large number of British agricultural farms that the efficiency distribution mostly follows either a J-shaped or, a bimodal curve. An exponential density above or, a linear mixture of two exponentially distributed variates would naturally fit this pattern. Furthermore the exponential density for outputs in (6.6) provides a natural interpretation of its two parameters m_j, k_j. For a homogeneous cluster (or industry) three cases may be distinguished: (a) only m_j may vary across n DMUs, while k_j may be the same k_0 say for all j, (b) only k_j may vary across n units while $m_j = m_0$ for all j=1,2,..,n and (c) both k_j and m_j may vary. The sources of inefficiency may be located more precisely if these cases are distinguished and a stochastic production frontier interpretation is made. Following the interpretation of Aigner, Lovell and Schmidt [18] the two sources of inefficiency due to variations in k_j and m_j may be related to different aspects of heterogeneity in inter-firm distribution of efficiency.

3.2 Stability and Robustness Aspects

We have not so far discussed one of the basic problems in applying the DEA model in the form (4.1) by using the representative unit's inputs in the objective function, which avoids the problem of computing n LP models, one for each reference unit. This problem arises when for each unit j we have more than one observation as samples thus giving rise to an enlarged data set on the inputs and outputs. In case of one output and m inputs for each DMU, let T_j denote the set of all possible input output combinations (x_j, y_j) where it is assumed that the vectors x_j and scalars y_j have all positive components. Thus the set T_j may be termed the technology set containing many feasible activities like (x_j, y_j) for each $j=1,2,..,n$. Since it is a production set and no joint production is permitted, one may reasonably assume each T_j to be convex. Now we seek the conditions on the enlarged data set $(X, y, \bar{x})$ such that the optimal vector β^* would remain invariant. Two cases are possible. One arises when the data are all realized and hence nonstochastic and the second allows stochastic variations. We first consider the deterministic case and define the feasible region for the output vector as

$$R(X,y) = \{y: X'\beta \geq y; y \geq 0; \beta \geq 0\} \tag{7.1}$$

The complete feasible region for the output vector y in (7.1) is then the union of all sets $R(X,y)$ for $(x_j, y_j) \varepsilon T_j$ for $j=1,2,\ldots,n$. Now we define a "dominant technology" by the set (X^*, y^*) i.e. a set of activities (x_1^*, y_1^*), (x_2^*, y_2^*), ..., (x_n^*, y_n^*) such that all feasible regions $R(X,y)$ for $(x_j, y_j) \varepsilon T_j$, $(j=1,2,\ldots,n)$ are contained in $R(X^*, y^*)$, where $R(X^*, y^*)$ is the feasible region for the technology (X^*, y^*). This concept of dominant technology is due to Johansen [12], who applied it to explain nonsubstitution in the Leontief-type input-output models. It is clear that a dominant technology (X^*, y^*) may not always exist for a given cluster. But if it does, then it has two important implications. One is that the complete feasible region for output y is the same as $R(X^*, y^*)$ and hence any attainable y-vector can be produced by employing the activity set $(x_1^*, y_1^*), (x_2^*, y_2^*), \ldots, (x_n^*, y_n^*)$. Secondly, even when there are feasible activities (X,y) other than (X^*, y^*), there is no need to substitute them since it would not increase the efficiency of the industry or the cluster.

Theorem 3

If there exist a vector $\lambda^*=(\lambda_1^*,\lambda_2^*,\ldots,\lambda_n^*)$ of nonnegative prices and a collection of activities $(x_j^*,y_j^*)\varepsilon T_j (j=1,2,\ldots,n)$ which satisfy the two conditions

(efficiency) $\quad \bar{x}_i - \sum_{j=1}^{n} x_{ij}\lambda_j^* \geq \bar{x}_i^* - \sum_{j=1}^{n} x_{ij}^*\lambda_j^* \quad$ (7.2)

$$\text{for all } (x_j,y_j)\varepsilon T_j(j=1,2,\ldots,n) \text{ and } i=1,2,\ldots,m$$

and

(zero slack) $\quad \bar{x}_i^* - \sum_{j=1}^{n} x_{ij}^*\lambda_j^* = 0 \quad \begin{array}{l}(j=1,2,\ldots,n)\\(i=1,2,\ldots,m)\end{array} \quad$ (7.3)

then (X^*,y^*) which comprises the collection $(x_1^*,y_1^*), (x_2^*,y_2^*),\ldots,(x_n^*,y_n^*)$ of activities is a dominant technology.

Proof

Consider any $\hat{y}_j$ belonging to $R(X,y)$. It must satisfy for all $j=1,2,\ldots,n$ for some $\beta \geq 0$:

$$\hat{y}_j \leq \sum_{i=1}^{m} \beta_i x_{ij}; \quad j=1,2,\ldots,n$$

If $\hat{y}_j$ is on the efficiency frontier we must also have $\hat{y}_j = \sum_{i=1}^{m} \beta_i^* x_{ij}^*$ for some $\beta^* \geq 0$ hence

$$\sum_{i=1}^{m} \beta_i x_{ij} \geq \sum_{i=1}^{m} \beta_i^* x_{ij}^* \quad (7.4)$$

CHAPTER VII

Next we multiply both sides of the efficiency conditon (7.2) by the nonnegative price β_i and sum over i to get

$$\sum_{ij}\beta_i x_{ij}\lambda_j^* - \sum_i \beta_i \bar{x}_i \leq \sum_{ij}\beta_i x_{ij}^* \lambda_j^* - \sum_i \beta_i \bar{x}_i^*$$

$$= \sum_i \beta_i [\sum_j x_{ij}^* \lambda_j^* - \bar{x}_i^*]$$

The last term on the right-hand side is zero by the zero-slack condition (7.3); hence

$$\sum_{ij}\beta_i x_{ij}\lambda_j^* \leq \sum_i \beta_i \bar{x}_i \qquad (7.5)$$

Multiply both sides of (7.4) by λ_j^* and sum over j to obtain:

$$\sum_{ij}\beta_i^* x_i^* \lambda_j^* \leq \sum_{ij}\beta_i x_{ij}\lambda_j^* \leq \sum_i \beta_i \bar{x}_i$$

where (7.5) has been used in the last step. Hence we obtain

$$\sum_{i=1}^m \beta_i^* \bar{x}_i^* \leq \sum_{i=1}^m \beta_i \bar{x}_i \qquad (7.6)$$

where $C^* = \sum_{i=1}^m \beta_i^* \bar{x}_i^*$ defines the minimum cost, which also equals the maximal value of output $V^* = \sum_{j=1}^n \lambda_j^* y_j^*$. Since the minimum cost technology (X^*, y^*) always satisfies the condition (7.6), there will in effect be no substitution for other technologies (X, y) so long as they satisfy (7.6).

Remark 3.1

If the prices (β_i^*, λ_j^*) are all positive for $i,j=1,2,..,m$ and $m<n$ and unique, then it is easy to show that (7.6) would hold as strict inequality i.e.

$$\sum_{i=1}^{m} \beta_i^* \bar{x}_i^* < \sum_{i=1}^{m} \beta_i \bar{x}_i$$

Remark 3.2

If $\bar{x}_i$ is identical with $\bar{x}_i^*$ (i.e. mean-preserving data variations), then the efficiency condition (7.2) implies that any input i belonging to the dominant technology is more valuable than the competing technology, provided the optimal output prices $\lambda^*=(\lambda_j^*)$ are used in both cases.

As a simple example of dominant technology consider the data set used by Charnes and Cooper [2] which represent a situation in which three DMUs produce one unit of output by using two types of inputs

		DMU's	
Inputs	1	2	3
x_1	2	$3+\varepsilon$	4
x_2	2	2	1

where a term $\varepsilon \geq 0$ has been added to indicate data variations. On using an objective function $C=\beta_1 \bar{x}_1 + \beta_2 \bar{x}_2 = 3\beta_1 + 1.67\beta_2$. We obtain the optimal solution $\beta_1^* = 1/6$, $\beta_2^* = 1/3$ when ε is set to zero. The optimal values of λ^* are $\lambda_1^*=0.61$, $\lambda_2^*=0.0$ and $\lambda_3^*=0.44$. Now for any data variations in the form of adding the nonnegative term ε would not alter the dominant technology specified by the input vectors of first and third DMUs. We will consider a more general application later.

Next we consider the case of stochastic variations in each of the variables x_{ij}, $\bar{x}_i = m_i$ and y_j. The sample observations are denoted by $x_{ij}(w)$, $m_i(w)$ and $y_j(w)$ where $w=1,2,\ldots,N$ is the number of observations. The sample means over the N observations are $\bar{x}_{ij}$, $\bar{m}_i$ and $\bar{y}_j$, for which the corresponding population parameters are denoted by μ_{ij}, μ_i and θ_j respectively. We assume for simplicity that the sample means converge to the population parameters as $N \to \infty$. Then one can establish a result analogous to that of Theorem 3.

Theorem 4

If there exist a vector $\bar{\lambda}^*$ of nonnegative prices and a collection of activities $(\bar{x}_j^*, \bar{y}_j^*) \varepsilon T_j$ $(j=1,2,\ldots,n)$ which satisfy the two conditions of efficiency and zero-slack (7.2) and (7.3), then $(\bar{X}^*, \bar{y}^*)$ which comprises the collection $(\bar{x}_1^*, \bar{y}_1^*)$, $(\bar{x}_2^*, \bar{y}_2^*)$, $\ldots$, $(\bar{x}_n^*, \bar{y}_n^*)$ of activities is a dominant technology in the mean. Furthermore, as $N \to \infty$, the dominant technology in the sample mean approaches its population counterpart (μ^*, θ^*) with probability one, where $\mu^* = (\mu_{ij}^*)$, $\theta^* = (\theta_j^*)$.

Remark 4.1

If the condition of a dominant technology set are satisfied by data variations, then it specifies an aspect of robustness since it excludes any substitution.

Remark 4.2

If the input-output data have time trends and different DMUs experience heterogeneity in growth, the dominant technology set in the mean may not appropriately capture the dynamic efficiency aspect.

3.3 Dynamic Aspect

The DEA model does not distinguish between the capital inputs which are durable and the current inputs which are not. In many production problems in economics however such a distinction is very important. Hence we may now consider a dynamic version of the DEA model (4.1) when there are two groups of inputs: current inputs which are not durable and capital units which are. If the first m_1 inputs are current and the remaining $(m-m_1)$ are capital inputs which depreciate at the rate d_{ij}, $0 < d_{ij} < 1$ per year, then the DEA model (4.1) would be transformed as follows

$$\min_{\beta \geq 0} \bar{g} = \sum_{i=1}^{m_1} \beta_i \bar{x}_i + \sum_{i=m_1+1}^{m} [(\frac{1}{n}) \sum_{j=1}^{n} \beta_i x_{ij}(1-d_{ij})]$$

subject to

$$\sum_{i=1}^{m_1} \beta_i x_{ij} + \sum_{i=m_1+1}^{m} \beta_i x_{ij}(1-d_{ij}) \geq y_j$$

$$j=1,2,\ldots,n$$

If the depreciation rates are known constants and given by the matrix $D=(d_{ij})$, one could solve for an optimal $\beta^* = \beta^*(D)$ and compare it with the case when there is no depreciation i.e. $d_{ij}=0$ for all i and j. Likewise the phenomenon of "learning by doing" or, capital-augmenting technological improvement can be easily introduced by replacing each capital input $x_{ij}(t)$ by $(1+u_{ij}(t))x_{ij}(t)$ for $i=m_1+1,\ldots,m$, where $u_{ij}(t) \geq 0$ represents productivity improvement. If these utilization rates depend on the level of inputs applied, the DEA model would be nonlinear

$$\min \bar{g} = \sum_{i=1}^{m_1} \beta_i \bar{x}_i + \frac{1}{n} \sum_{i=m_1+1}^{m} [y_i \sum_{j=1}^{n} x_{ij}(1+u_{ij}(x))]$$

subject to

$$\sum_{i=1}^{m_1} \beta_i x_{ij} + \sum_{i+m_1+1}^{m} \gamma_i x_{ij}(1+u_{ij}(x)) \geq y_j(t) + \Delta y_j$$

$$\beta_i \geq 0, \gamma_i \geq 0; \quad j=1,2,\ldots,n$$

where Δy_j denotes a nonnegative increase of output of j-th DMU. If any input could be used partly as current inputs and partly as investment, where the latter may lead to increased outputs in later time periods, then a simple dynamic model analogous to (4.1) can be formalized as follows:

$$\min_{\beta,\gamma} \bar{g} = \sum_{i=1}^{m} \beta_i \bar{x}_i + \sum_{i=1}^{m} \gamma_i \overline{\Delta x}_i$$

subject to (8.1)

$$\sum_{i=1}^{m} \beta_i x_{ij}(t) + \sum_{i=1}^{m} \gamma_i \Delta x_{ij} \geq y_j(t) + \Delta y_j$$

$$\beta_i \geq 0; \gamma_i \geq 0; \quad i=1,2,\ldots,m; \quad j=1,2,\ldots,n$$

Here $\overline{\Delta x}_i$ is the mean level of the incremental inputs i.e., $\overline{\Delta x}_i = (1/n) \sum_{j=1}^{n} \Delta x_{ij}$. If both $x_{ij}(t)$ and Δx_{ij} are observable for the n DMUs, along with their outputs y_j and Δy_i, one could apply DEA model as above to solve for the optimal weights (β^*, γ^*). Note that the optimal weights β^*, γ^* are interdependent through the allocation of inputs x_{ij} between current and future uses. In the static version of the model, Δx_{ij} and Δy_j would reduce to zero as in the LP model (4.1). Thus if a particular DMU turns out to be different in the static framework, one may have to look forward to its capital-augmenting allocation of the inputs to check if it is efficient in an intertemporal sense. Once the relative inefficiencies (or inefficient units) are identified by the dynamic LP model

(8.1), one has to search for policies to improve efficiency. In some cases when the incremental inputs or outputs have longer gestation lags e.g., $\Delta x_{ij} = x_{ij}(t+k) - x_{ij}(t)$, and $\Delta y_j = y_j(t+k) - y_j(t)$, $k \geq 2$ due to the introduction of a new technology say, then it may be reasonable to decompose the above model (8.1) into two parts e.g.,

$$\min \bar{g}(\beta) = \Sigma_i \beta_i \bar{x}_i \qquad \min \bar{g}(\gamma) = \Sigma_i \gamma_i \overline{\Delta x}_i$$

and

$$\text{subject to } \Sigma_i \beta_i x_{ij} \geq y_j \qquad \text{subject to } \Sigma_i \gamma_i \Delta x_{ij} \geq \Delta y_j$$

$$\beta_i \geq 0 \qquad \gamma_i \geq 0$$

and analyze the efficiency weights β, γ separately to see how the relative efficiency of an individual DMU is changing in relation to other DMUs in the overall system. The optimal efficiency prices are now decomposed into two parts β^* and γ^* with their corresponding dual vectors λ^* and μ^* say.

These dynamic versions of the DEA model require that we have observed the use of both current and capital inputs and the realized outputs. To apply this in a planning sense with a future horizon from $t=1$ to $t=T$ we have to use expected inputs and outputs. On using these "expectational" or planned variables one could define a dynamic scale of measurement of expected efficiency. For this purpose let us assume that each input $x_{ij}(t)$ can be decomposed into three parts i.e. $x_{ij}(t) = x_{ij}^C(t) + x_{ij}^K(t) + x_{ij}^D(t+1)$ where $x_{ij}^C(t)$ are current inputs, $x_{ij}^K(t)$ are capital inputs at the beginning of period t and $x_{ij}^D(t+1)$ are the depreciated capital inputs available to j-th DMU at the beginning of the next period. The dynamic version of the DEA model may then be formulated as follows:

$$\text{Min } C(\beta,\gamma) = \sum_{t=1}^{T} r^t [\sum_{i=1}^{m} \{\beta_i(t)\bar{x}_i^C(t) + \gamma_i(t)\bar{x}_i^K(t)$$
$$+ \gamma_i \bar{x}_i^D(t-1)\}]$$

subject to (8.3)

$$\sum_{i=1}^{m} \{\beta_i(t)x_{ij}^C(t) + \gamma_i(t)x_{ij}^K(t)\} \geq y_j(t) + \sum_{i=1}^{m} \gamma_i(t)x_{ij}^D(t-1)$$

$$j=1,2,\ldots,n; \quad \beta_i(t) \geq 0, \quad \gamma_i(t) \geq 0, \quad i=1,2,\ldots,m$$

where $0 < r < 1$ is the constant rate of discount and the mean input $\bar{x}_i$ is decomposed into three component inputs $\bar{x}_i^C$, $\bar{x}_i^K$ and $\bar{x}_i^D$. This dynamic model (8.3) may now be interpreted in two ways according as the planned quantities x_{ij}^C, x_{ij}^K and x_{ij}^D are assumed to be realized or not. If they are assumed realized, it would generate a set of optimal values of the vectors $\beta^*(t)$ and $\gamma^*(t)$ and the associated dual vectors $\lambda^*(t)$, which can be used to characterize efficiency distribution as before for each current t and over the planning horizon. In the second case the planned quantities may differ from the realized levels, thus making realized prices $\beta(t)$, $\gamma(t)$ and $\lambda(t)$ differ from their optimal values. In this case we have the dynamic inconsistency problem that has been widely discussed in dyanmic policy models under uncertainty [19].

In case the planned input and output operations are realized, the optimal vectors $\beta^*(t)$, $\gamma^*(t)$ and $\lambda^*(t)$ must satisfy for each t ($t=1,2,\ldots,T-1$) the following necessary conditions

$$\sum_{j=1}^{n} \lambda_j(t) x_{ij}^C(t) \leq \bar{x}_i^C(t); \quad i=1,2,\ldots,m \qquad (8.4)$$

$$\sum_{j=1}^{n} [r\lambda_j(t+1) x_{ij}^D(t+1) - \lambda_j(t) x_{ij}^K(t)]$$

$$\geq r\bar{x}_i^D(t+1) - \bar{x}_i^K(t); \quad i=1,2,\ldots,m$$

$$\lambda_j(t) \geq 0, \quad j=1,2,\ldots,n \qquad (8.5)$$

It is clear that if the depreciated inputs $x_{ij}^D(t+1)$ at the beginning of period t+1 (i.e. left over from period t) are of no value i.e. of zero salvage value, then the constraint (8.5) for capital reduces to

$$\sum_{j=1}^{n} \lambda_j(t) x_{ij}^K(t) \leq \bar{x}_i^K(t)$$

This is exactly of the same form as the current inputs. In case of positive salvage value and an optimal interior solution, the optimal quantities $\lambda_j^*(t)$ will satisfy the linear difference equation

$$\sum_{j=1}^{n} [r\lambda_j^*(t+1) x_{ij}^D(t+1) - \lambda_j^*(t) x_{ij}^K(t)]$$

$$= r\bar{x}_i^D(t+1) - \bar{x}_i^K(t)$$

along the optimal trajectory. If the steady state of this difference equation in vector $\lambda^*(t)$ belongs to the optimal trajectory, one could evaluate the process of efficiency distribution over time. In particular this dynamic version of the DEA model helps to generalize it in three important directions.

First, the myopic policy of efficiency improvement by ignoring all future investment and its output-augmenting effects may be compared and contrasted with that of an intertemporal optimal policy over the given planning horizon. Second, the planned quantities of the three inputs $x_{ij}^C(t)$, $x_{ij}^K(t)$ and $x_{ij}^D(t+1)$ may be related to their contributions to total current value of optimal output $\sum_{j=1}^n y_j^*(t) y_j(t)$ thus making the efficiency ranking process more precise and useful. In particular one may search for positive slack in $x_{ij}^D(t+1)$ for the relatively inefficient units, which may not yet have utilized their durable capital inputs to the fullest extent. Lastly, the efficiency measure in its current and intertemporal aspects can be clearly identified in terms of its sensitivity to the rate of discount (r), and the implicit policy followed for depreciating capital inputs. However, implementing such a dynamic model sets up heavy demands on comprehensive data both realized and expected.

4. An Empirical Application

The illustrative empirical application considers an urban school district with 24 DMUs, where the single output variable (y_j) is the achievement scores of sixth grade public elementary school pupils and the four inputs selected are teacher salary (x_1), average class size (x_2), proportion of minority students (x_3) and a proxy variable (x_4) indicating students' parental background. Three types of tests are performed to test the sensitivity and robustness of the optimal coefficients β^* and λ^* defined before in the DEA models (4.1) and (4.2) respectively.

The data are from selected public elementary schools in California for the year 1977-78, over which regression estimates of production functions have been made in earlier studies [20]. Both inputs and outputs are in logarithmic units and clearly some of the inputs above have no obvious market prices. The sample set of 25 units were selected from a larger set of 50 to retain more homogeneity and the LP model (4.1) is then run to determine the 100% efficient units and order all the units in a decreasing order of efficiency. The sample set is then divided into 6 groups with 4 in each of first five groups, such that the first group contained the 100% efficient units, the second group contained the next 4 efficient (but less than 100%) units and so on, till the sixth group contained the last 5 units. The mean input vectors of these 6 groups are $\bar{x}_{(1)}$ through $\bar{x}_{(6)}$ respectively. Six LP models from LP(1) through LP(6) are run with these mean input vectors in the objective function of the model (4.1). This is referred to as Test 1, which provides an analsis of the sensitivity of the optimal β^* coefficients. For Test 2, 4 units from the top are successively deleted from the constraints and the mean input levels (these deletions are done up to n=12) in the objective function, thus generating four LP models with n=25, 21, 17 and 13. This test shows the effect of truncation in the sample space on the optimal coefficients β^* which characterize the production frontier.

For Test 3, we use the LP model (3.1) where each DMU is represented in the objective function instead of the mean representative unit. Thus we have n LP models. The DMUs are marked by serial numbers 2 through 26, where the mean DMU is given the number 1. The objective of this test is to determine which DMUs had the highest frequency of reaching the 100% efficiency level in the 25 LP models from LP_1 through LP_{25}. Unlike the first two tests, this Test 3 provides some insight into the connections of the LP approach of efficiency with the statistical regression approach of an average production function.

The results of Test 1 are as follows, where LP(0) denotes the case of the representative DMU with mean inputs in the objective function:

	LP(0)	LP(1)	LP(2)	LP(3)	LP(4)	LP(5)	LP(6)
β_1^*	0.247	0.247	0.247	0.247	0.451	0.0	0.359
β_2^*	0.626	0.626	0.626	0.626	0.0	1.503	0.350
β_3^*	0.523	0.523	0.523	0.523	0.206	1.169	0.146
β_4^*	0.179	0.179	0.179	0.179	0.074	0.417	0.079
max λ_j^*	0.250	0.250	0.392	0.397	0.958	0.550	0.601

It is clear that the β^* coefficients for the three LP models LP(1) through LP(3) are insensitive to data variations affecting the objective function, although the latter are widely dispersed between the efficient and the less efficient units. These cases are also found to satisfy the conditions of Theorem 1 in terms of their closeness to the representative unit denoted by LP(0). This may be more apparent when one looks at the overall frequency distribution of the efficiency ratio e_j defined as $e_j = 1-(y_j/y_j^*)$

domain of e_j:	0.0-0.04	0.05-0.07	0.08-0.11	0.12-0.15	above 0.15
% frequency:	36	20	28	4	12

The distribution of the efficiency ratio e_j is clearly seen to be nonnormal and also asymmetrical. Note the high frequency of units in the efficiency range $y_j/y_j^* \geq 0.97$, from which it is clear that a deterministic DEA model would not include all the top 36% of the DMUs very near the 100% efficiency level.

To test the form of the probability distribution of the efficiency ratio e_j defined above we applied the nonparametric Kolmogorov-Smirnov statistic as follows.

Let the cumulative distribution of efficiency e_j in a cluster be denoted by $F(e)$ and it is assumed to be continuous; let $F_N(e)$ be the empirical distribution based on N mutually independent trials of the random variable e (and

EFFICIENCY MEASUREMENT IN NONMARKET SYSTEMS 363

constructed as cumulative frequency polygons from ordered values of observed e). Then, by using the Kolmogorov-Smirnov theory one could derive the following probabilistic relation:

$$\text{Prob}\left[\sup_{0 < F(e) < 1} \frac{F_N(e)}{F(e)} < t\right] = h(t) = \begin{cases} 0, & \text{for } t \leq 1 \\ 1 - \frac{1}{t}, & \text{for } t > 1 \end{cases} \quad (8.6)$$

which is independent of the form of the parent population, except that it must be continuous. This relation (8.6) can be easily derived. For example, let $t \geq 1$ and let ε_k denote the largest root of the equation

$$F(e) = \frac{k}{Nt}, \quad 1 \leq k \leq N \quad (8.7)$$

which exists since it is assumed that $F(e)$ is continuous. If the probability of the event on the left-hand side of (8.6) is not zero and by construction $F_N(e)$ is nondecreasing with the ratio $F_N(e)/F(e)$ tending to one as $N \to \infty$, we must have a value e_0 such that

$$\frac{F_N(e_0)}{F(e_0)} = t$$

But by construction $F_N(e_0) = k/N$, k is a positive integer; hence e_0 may be taken as a special value of ε_k so that the inequality

$$e_k^* < \varepsilon_k < e_{k+1}^*$$

holds. By generating a sequence of ε_k values from the cumulative frequency polygons which are constructed from observed values of e_j arranged in an ascending order, one could prove the result (8.6).

Two advantages of this result are that it is nonparametric and hence applicable rather widely and, if the cluster contains non-efficient units it would be detected for some interval t at which the probability h(t) would be positive in (8.6). In terms of the distribution of h(t) values, one may note how the distribution changes as N increases, or in some cases two or more similar clusters may be compared for their divergence. These results may also be compared with parametric forms of specified populations, from which observations are assumed to be generated. Two parameteric population distributions were hypothesized: normal and exponential. Whereas the normal distribution was rejected as the null hypothesis by our data set, the two-parameter exponential density was not.

The optimal β^* values for Test 2 show that some β^*_j coefficients are

Results of Test 2

n	β^*_1	β^*_2	β^*_3	β^*_4
25	0.247	0.626	0.523	0.179
21	0.151	0.935	0.804	0.276
17	0.100	1.087	0.682	0.241
13	0.175	0.692	0.701	0.189
Regression	0.270**	0.861**	0.971**	0.320** $R^2=0.77$

more sensitive to truncation in the sample space than others. For comparison we have also reproduced in the last row the ordinary linear regression estimates based on n=25 observations. The double asterisks indicate that the estimated coefficients are statistically significant at 1% level of the t-test. Except for the β^*_4 value, the LP estimates of β^* are usually lower than the regression estimates for n=25. Truncation has the general tendency to increase the value of the individual β_i^* coefficients estimated by the LP model but this may not hold for each individual coefficient.

The objective of Test 3 is to determine which of the 25 DMUs had the highest frequency of attaining the 100% efficiency level under any of the 25 objective functions, one for each DMU. For this purpose we had to distinguish between the degenerate and nondegenerate LPs:

	Serial Numbers of DMUs							
	Nondegenerate LPs(n=11)				Degenerate LPs(n=14)			
	24	17	4	26	24	17	4	26	8
% frequency	100	82	54	54	78	21	50	28	28

It is clear that in terms of the highest frequency or mode criterion, DMU_{24} is the most efficient, followed by DMU_{17} and DMU_4. Three linear regression functions were estimated with output regressed on the four inputs as follows: Case I included only the data of the nondegenerate LPs(n=11) above, Case II included only the data from the degenerate LP models and Case III contained data for all n=25 units. The R^2 values turned out to be

	Case I	Case II	Case III
R^2	0.96	0.79	0.77

It is apparent that the LP models of the DEA model can be most profitably used in the nondegenerate cases as a screening device to reduce the heterogeneity of the observed data set and to improve the predictive power of the regresson model. In this sense the dummy variable approach of the conventional regression model may be viewed as a special case of this screening device suggested by the DEA model. More simulation work in cases involving various nonnormal distributions of efficiency would be very helpful here.

CHAPTER VII

References

1. Charnes, A., Cooper, W.S. and E. Rhodes. Measuring the efficiency of decision making units. European Journal of Operational Research 2(1978), 429-444.

2. Charnes, A. and Cooper, W.W. Management science relations for evaluation and management accountability. Journal of Enterprise Management 2(1980), 143-162.

3. Banker, R.D., Charnes, A., Cooper, W.W. and Schinnar, A.P. A bi-extremal principle for frontier estimation and efficiency evaluation. Management Science 27(1981), 1370-1382.

4. Farrell, M.J. and Fieldhouse, M. Estimating efficiency in production functions under increasing returns to scale. Journal of Royal Statistical Society. Series A 125(1962), 252-267.

5. Bowlin, W.F., Charnes, A., Cooper, W.W. and Sherman, H.D. Data envelopment analysis and regression approaches to efficiency estimation and evaluation. Paper presented at Fourth Annual Conference on Current Issues in Productivity held at Cornell University, Ithaca, New York, November 30, 1982.

6. Debreu, G. The coefficient of resource utilization. Econometrica 49(1951), 273-292.

7. Diewert, W.E. The measurement of waste within the production sector of an open economy. Scandinavian Journal of Economics 85(1983), 159-179.

8. Hanoch, G. and Rothschild. Testing the assumptions of production theory: a nonparametric approach. Journal of Political Economy 80(1972), 256-275.

9. Diewert, W.E. and Parkan, C. Linear programming tests of regularity conditions for production function, in Quantitative Studies on Production and Prices. Wurzburg, Austria: Physica-Verlag, 1983.

10. Sengupta, J.K. and Sfeir, R.E. Production frontier estimates of scale in public schools in California. Economics of Education Review 5, No. 3, 1986.

11. Afriat, S.N. Efficiency estimation of production functions. International Economic Review 13(1972), 568-598.

12. Johansen, L. Simple and general nonsubstitution theorems for input-output models. Journal of Economic Theory 5(1972), 383-394.

13. Timmer, C.P. Using a probabilstic frontier production function to measure technical efficiency. Journal of Political Economy 79(1971), 776-794.

14. Aigner, D.J. and Chu, S.F. On estimating the industry production function. American Economic Review 58(1968), 826-839

15. Nunamaker, T.R. Using data envelopment analysis to measure the efficiency of monprofit organizations: a critical evaluation. Managerial and Decision Economics 6(1985), 50-58.

16. Peleg, B. and Yaari, M.E. A price characterization of efficient random variables. Econometrica 43 (1975), 283-292 (1975).

17. J.K. Sengupta. Decision Models in Stochastic Programming. North Holland, New York, 1982.

18. Aigner, D., Lovell, C. and Schmidt, P. Formulation and estimation of stochastic frontier production function models. Journal of Econometrics 6(1977), 21-37.

19. Sengupta, J.K. Information and efficiency in economic decision. Dordrecht: Martinus Nijhoff Publishers, 1985, ch. 12.

20. Sengupta, J.K. Production frontier estimation to measure efficiency: a critical evaluation in the light of data envelopment analysis. To appear in: Managerial and Decision Economics, 1986.

Author Index

Afriat, S. N., 366

Ahmed, N. U., 115, 122

Aigner, D. J., 335, 366

Amemiya, T., 171, 189

Andersen, P., 35, 51

Arditti, F. D., 229, 232, 319

Astrom, K. J., 10, 101, 121

Aumann, R. J., 60, 120

Baba, N., 108, 122

Banker, R. D., 335, 366

Bawa, V. S., 128, 188, 319

Beedles, W. L., 129, 188, 319

Berman, O., 232, 320

Bertsekas, D. P., 178, 189

Bharucha-Reid, A. T., 120

Black, F., 130, 188

Bresnahan, T. F., 120

Brock, W. A., 51, 78, 121

Brown, S. J., 128, 188

Calmfors, L., 121

Campanella, F. B., 275, 319

Cauwenberghe, A. R., 101, 121

Charnes, A., 5, 10, 321, 335, 366

Chen, Y. C., 115, 122

Chu, S. F., 335, 366

Clark, C., 110, 122

Clarke, D. W., 101, 121

Cooper, W. W., 5, 10, 321, 335, 366

Copeland, T. E., 232, 287, 319

Cornell, B., 287, 320

Danskin, J. M., 51

Davis, M. H. A., 10

Debreu, G., 323, 366

Diewert, W. E., 10, 323, 366

Dixit, A., 54, 120

Elton, E. J., 319

Farrell, M. J., 334, 366

Fershtman, C., 54, 120

Fieldhouse, M., 334, 366

Gawthrop, P. J., 103, 121

Geer, T., 91, 121

Ghosh, J. K., 189

Gibbons, J. D., 271, 320

Gleit, A., 104, 122

Goodwin, G. C., 32, 51

Goreux, L. M., 84, 121

Grossman, S., 287, 320

Gruber, M. J., 319

AUTHOR INDEX

Hanoch, G., 366

Harsanyi, J. C., 58, 120

Ho, Y. C., 120

Huber, P. J., 224, 232

Hughes-Hallett, A. J., 84, 121

Ito, 19

Jacob, N. L., 143, 188, 319

Jensen, M. C., 232, 285, 320

Johansen, L., 3, 10, 350, 366

Kailath, T., 51

Kallianpur, G., 189

Kamien, M. I., 54, 120

Karlin, S., 178, 189

Keyser, R. M. C., 101, 121

Klein, R. W., 128, 188, 319

Kreps, D. M., 53, 120

Krishnaiah, P. R., 189

Lehmann, E. L., 139, 169, 188

Levy, H., 229, 232, 309, 320

Lewis, T. R., 51

Lovell, C., 367

Ljung, L., 101

Mains, N. E., 286, 319

Malliaris, A. G., 51

Mangel, M., 14, 51, 122

Mendelssohn, R., 104, 122

Milgrom, P., 53, 120

Modiano, E., 232, 320

Monopoli, R. V., 121

Mori, T., 52

Moulin, H., 61, 120

Narendra, K. S., 121

Newbery, D. M. G., 83, 121

Neyman, J., 170, 188

Nguyen, D. T., 84, 121

Noldus, E., 52

Nunamaker, T. R., 335, 367

Olkin, I., 271, 320

Parkan, C., 10, 324, 366

Parlar, M., 51

Payne, R. L., 32, 51

Peleg, B., 342, 367

Phillips, A. W., 49, 51

Pindyck, R. S., 51, 110, 122

Restrepo, R. A., 120

Reynolds, P. D., 88, 121

Rhodes, E., 366

Roberts, J., 53, 120

Roll, R., 232, 237, 287, 320

Rothschild, M., 366

Sarnat, M., 232, 319

Schinnar, A. P., 335, 366

Schmidt, P., 367

Schnabel, J. A., 232, 320

Scholes, M., 130, 188

Schwartz, N. L., 54, 120

Scott, E. L., 170, 188

Sengupta, J. K., 10, 14, 51, 68, 120, 188, 232, 367

Sobel, M., 271, 320

Soong, T. T., 43, 52

Spence, A. M., 53, 120

Starr, A. W., 120

Stenlund, H., 100, 121

Stone, L. D., 52

Szego, G. P., 125, 188, 232, 319

Taylor, J. B., 100, 121

Timmer, C. P., 335, 366

Tintner, G., 14, 51, 120, 320

Tobin, J., 188

Toutenberg, H. 214, 232, 320

Vinter, R. B., 10

Weston, J. F., 232, 287, 319

Whittle, P., 84, 121

Wittenmark, B., 10, 101, 121

Yaari, M. E., 342, 367

Subject Index

Adaptive behavior, 42, 97

Adaptive control, 1, 30

Arrow-Pratt risk aversion, 342

Asymptotic convergence, 59

Automata theory, 107

Average production function, 170

Bargaining solution, 3

Bayesian method, 42, 140

Birth and death process, 23

Brownian motion, 16, 22

Buffer stock program, 94

Causal system, 102

Cautious policy, 98

Certainty equivalence, 81, 184

Chance-constrained programming, 344

Chapman-Kolmogorov equation, 74

Characteristic equation, 92

Closed loop system, 55

Conjectural equilibria, 2, 53

Correlated strategies, 62

Cournot model, 40, 82

Cournot-Nash equilibrium, 60, 72

Data envelopment analysis, 5, 321

Decision rule, 184

Differential game, 2, 58

Diffusion process, 33

D-partition method, 92

Diversification of investment, 123

Dynamic programming, 105

Econometric tests of portfolio, 293

Efficiency evaluation, 325

Efficiency frontier, 152

Efficiency measure, 321

Eigenvalue problem, 73

Expected loss, 99

Feedback policy, 55

Fokker-Planc equation, 44

Gaussian process, 17

Hamiltonian, 28

Inconsistency problem, 2

Independent increment, 16

Integrated variance test, 249

Ito's rules, 19

SUBJECT INDEX

Jump process, 23
Kalman-Bucy filtering, 183
Kalman filter, 29, 31
Keynesian model, 20
Kolmogorov-Smirnov statistic, 362
Kuhn-Tucker theory, 35
Lagrangian function, 151, 181
Learning model, 103
Least squares method, 30
Limit pricing, 54
Limited diversification portfolio, 143
Linear decision rule, 184
Linear programming, 334
Logistic function, 117
LQG model, 2
Lyapunov theorem, 80
Markov process, 15
Markowitz model, 193
Maximum likelihood method, 159
Maximum principle, 28
Mean variance models, 166
Method of moments, 47
Minimax solution, 143, 151, 168
Minimum variance portfolio, 127

Mixed strategy, 59
Monte Carlo simulation, 100
Multivariate distance, 336
Mutual funds, 235
Nash equilibrium, 55
Non-cooperative game, 2
Nonlinear problems, 5
Non-normality, 177
Nonparametric measure, 5
Non-zero sum game, 52
Normal distribution, 141
Open loop control, 55
Optimal control, 53
Optimal search, 38
Orthogonal portfolio, 190
Pareto efficiency, 321
Pay-off function, 81
Perron-Frobenius theorem, 149
Phillips model, 20
Pontryagin principle, 28
Portfolio theory, 123
Production frontier, 322
Production function, 322
Queueing process, 25
Quasi-polynomial, 50, 93
Random search, 3

SUBJECT INDEX

Rational expectations, 4
Rayleigh-Ritz algorithm, 61
Reaction function, 59
Regression model, 69, 102
Regulation problem, 98
Renewable resource model, 33, 110
Risk aversion, 6
Robust decision, 290, 341
Robustness property, 6, 154
Saddle point, 41
Search model, 3, 40
Self-tuning regulator, 100
Sequential updating, 98
Singular portfolio, 193
Stabilization policy, 83
Stationary process, 14
Stochastic control, 29
Stochastic game, 69
Stochastic optimization, 5
Stochastic process, 12
Systematic risk, 123
Trajectory, 186
Transition probability, 44
Transversality condition, 80
Two-stage problem, 30

Uncertainty in models, 8, 32
Unsystematic risk, 123
White noise, 18
Wiener process, 16, 43
Wiener-Levy process, 16
Wishart distribution, 159